Praise for *The Control Revolution*

"A must-read for anyone interested in computers and culture."
—*Technology Review*

"A balanced, carefully edited guide to the issues posed by the Internet."
—*The Washington Post*

"This is an extraordinarily powerful and mature story of the hopeful side of the Internet's revolution. Rich with insight, and surprisingly new conclusions, Shapiro's book is the best 'second-generation' thought on the questions the Net will raise. Beautifully written, and tightly argued, the book is certain to become a classic."
—**Lawrence Lessig, Berkman Professor, Harvard Law School and author of *Code***

"Illuminating, lucid, lively, and sensible. . . . The real value of these books [Shapiro's and Lessig's *Code*] lies less in their concrete recommendations than in their skepticism about libertarian platitudes, their insistence on the need for public attention to emerging problems, and their spirit of uncertainty and wariness about possible consequences of the Internet."
—**Cass Sunstein, *The New Republic***

"It's hard to imagine a more timely book about the real significance of the Internet."
—**Jon Katz, author of *Geeks*, on Slashdot.org**

The Control Revolution is bound to for years to come."
—**Steven Johnson, editor of *Feed* and auth**

"With scrupulous documentation a ...dgeable but unpatronizing tone, Shapiro delivers a penetrating analysis of both the promise and peril of the digital future."
—***Publisher's Weekly* starred review**

"Essential reading for those charged with creating the future." WITHDRAWN
—**Amazon.com**

"It's important for each of us to read *The Control Revolution*, and to take up [Shapiro's] call for a balance between personal and public interest."
—**Zoë Baird, President, The Markle Foundation**

"Put down your *Wired* magazine and pick up this book."
—**Borders.co**

the control revolution, _n._

1. the potentially monumental shift in control from institutions to individuals made possible by new technology such as the Internet;

2. the conflict over such change between individuals and powerful entities (governments, corporations, the media);

3. the unexpected, and not always desirable, ways in which such change could reshape our lives

The
Control
Revolution

How the
Internet
is Putting
Individuals
in Charge
and Changing
the World
We Know ▶ Andrew L. Shapiro

 A Century Foundation Book
PublicAffairs ▶ New York

The Century Foundation, formerly the Twentieth Century Fund, sponsors and supervises timely analyses of economic policy, foreign affairs, and domestic political issues. Not-for-profit and nonpartisan, it was founded in 1919 and endowed by Edward A. Filene.

Copyright © 1999 by Andrew L. Shapiro.

Published in the United States by PublicAffairs™, a member of the Perseus Books Group.

All rights reserved.

Printed in the United States of America.

No part of this book may be reproduced in any manner whatsoever without written permission except in the case of brief quotations embodied in critical articles and reviews. For information, address PublicAffairs, 250 West 57th Street, Suite 1321, New York, NY 10107.

Library of Congress Cataloging-in-Publication Data
Shapiro, Andrew L.
    The control revolution : how the internet is putting individuals in charge and changing the world we know / Andrew L. Shapiro.
        p.   cm.
    "A Century Foundation book."
    ISBN 1-891620-86-X
    1. Information society. 2. Information society—Political aspects. 3. Control (Psychology). 4. Power (Social sciences). I. Title.
    HM851.S53   1999
    303.48'33—dc21   99–24441
                                                                                          CIP

10 9 8 7 6 5 4 3 2

Contents

Foreword   *vii*

Introduction   *xi*

▶ **Part One: Revolution**

1 "We Have Revolution Now"   3

2 The Politics of Code   13

3 Gaining Control   25

4 Liebling's Revenge: The Power of Interactivity   34

5 Masters of Our Own Domains: Personalization
  of Experience   44

6 The Decline of Middlemen: Day Trading
  and "Electrified Voting"   53

▶ **Part Two: Resistance**

7 An Anxious State: Controlling Speech,
  Secrets, and Creativity   63

8 Where Do You Want to Go Today?:
  Microsoft and the Illusion of Control   84

▶ **Part Three: Oversteer**

9 Narrowing Our Horizons   105

10 A Fraying Net   115

11 Freedom from Speech   124

12 The Drudge Factor   133

13 Shopper's Heaven?   142

14 Push-button Politics   150

15 Privacy for Sale   158

▶ Part Four: Balance

16 Mapping Principles   169

17 Shattering Illusions   180

18 In Defense of Middlemen   187

19 In Defense of Accidents   197

20 Surf Globally, Network Locally   208

21 The Tools of Democracy   217

Epilogue: From Revolution to Resolution   231

Afterword   234

Notes   237

Bibliography   272

Acknowledgments   279

Index   281

Foreword

▶ For much of this century, the United States has been both the principal agent of the rapid and widespread dissemination of technical innovation and the nation most profoundly changed by it. In the field of communications, although other nations produce a fair share of the hardware, the U.S. continues to set the pace. The American approach to information, broadly conceived, is sweeping the globe at almost light speed, creating a revolution in how people see the world.

The changes under way go far deeper than Asians or Europeans acquiring a taste for McDonald's hamburgers or Michael Jordan's jump shot; they are bringing about a reconception of one of the most basic components of experience: time. We send the numbers NOW; we learn the world news NOW; we see the latest fashions NOW; we share the latest scandal NOW. We even have a new name for the sort of time in which these things happen and are shared with us: real time. The term suggests not just immediacy, but also, perhaps, a new sort of reality.

The twentieth century's information revolution has fueled an expanding set of expectations about what we can know as well as when we can know it. The evolutionary and transforming aspects of advanced communications are especially important because, in the modern world, the stream of data, news, gossip, comedy, sports, and other information seems to surround and permeate our lives. In this environment, defenders of local traditions and language everywhere feel overmatched. The process seems driven by an almost palpable and irresistible alien force—a force seemingly determined to assimilate all societies into a vast and uniform worldwide information culture. In fact, considering the pace of globalized communication, marketing, and commerce, it seems possible that resistance is futile.

Our daily lives were first drastically reordered by radio and then television. The spread of broadcasting, of course, knit diverse audiences together in ways that simply were previously impossible. And, more recently, in a process far from complete, the proliferation of specialized stations and channels, especially in the United States, has reversed the trend, fragmenting listeners and viewers into smaller, perhaps more homogeneous, groups. Strikingly, well before this shift from broadcasting to "narrowcasting" is anywhere near complete, yet another, potentially equally transforming, change in communications is occurring through the Internet.

In just the last five years, the Internet has provoked millions of words, a fair share of them superlatives. In fact, it apparently is impossible for anyone to write about the "Net" with restraint or understatement. At the same time, it is impossible to discuss the future effects of the Net with any degree of certainty. Like the stock of many companies with Internet businesses (or possible businesses), the long-term expectations about the Net keep escalating upwards in our imaginations. That is not to say that high expectations are unwarranted; it is, rather, to argue that we simply don't know yet what sort of expectations are rational. We need more information, more experience, and more time to think more about these issues. Fortunately, the author of this book, Andrew Shapiro, has done some good thinking for us.

Shapiro, director of the Aspen Institute Internet Policy Project, cuts through the uncertainties to explore some of the major issues that, given the nature of the Internet, are likely to grow in importance. His argument that new technology allows power to shift from the public to the private sector—and even down to the individual—is provocative and prophetic. Among other things it shows how the Internet may cause us to abandon the deliberation that is central to representative democracy.

Internet technology may produce analogous conflicts in other areas. Lots of information is available, for example, but its provenance and veracity is often unknown. Are most individuals actually in a position to judge the quality and importance of the stream of undigested and unchecked information that is available on the Net? We may know nothing concerning the reputation of the source of the information and be equally ignorant of the availability of other points of view.

So-called web browsers are merely that—they give us lots of choices but no wisdom. But we do not have to take advantage of the breadth of our new access to information. We can choose to connect to the world, so to speak, or choose to narrow our sources dramatically. Already, Internet users can personalize their news. If the practice becomes widespread and a common substitute for gathering information from other sources, it could have profound implications for public discourse, community, and democratic politics. The notion of small packets of information, exactly tailored to the specific interest of an individual sounds extremely attractive, with no downside, but only at first hearing. As Shapiro points out, if we choose only those with whom we agree as our reliable sources—whether our choice is the ACLU or the 700 Club—we may fundamentally narrow our perceptions of reality. He notes that such a process eliminates one essential premise of the search for truth: the importance of testing our ideas against competing arguments. Indeed, this sort of interchange is what makes our diverse republic work.

The study of the policy implications of the Internet is, like the medium itself, in its infancy. But, like television and radio before it, this youngster is growing up fast—and changing our lives in the process. Moreover, as Shapiro makes clear, the fundamental issues raised by the Internet regarding privacy, security, reliability, taste, and truth are complex and likely to become more so—and they are not likely to be resolved or even simplified any time soon.

The Century Foundation has a long history of examining the impact of technology on public issues—and vice versa. Over the past decade, for example, we published Lawrence Grossman's *The Electronic Republic*; *The New Information Infrastructure*, a volume of essays edited by William Drake; Jay Rosen and Paul Taylor's *The New News v. The Old News*; as well as two task force reports: *1-800-President: The Report of the Twentieth Century Fund Task Force on Television and the Campaign of 1992*; and *Quality Time: The Report of the Twentieth Century Fund Task Force on Public Television*. In addition, like other organizations, we have been surprised by the sudden importance of the Internet to our own activities. Within a few months of establishing our own web site, for example, we found that we were receiving more than 300,000 visits a month. Thus, Andrew Shapiro's pioneering examination of public policy in this area was of particular interest to us. His book fulfills our expectations and, we hope, will provoke more analyses of the effects of this potentially transforming force, known as the Internet, on life in America.

Richard C. Leone, President
The Century Foundation
March 1999

Introduction

▶ One of the curious things about living through a time of whirlwind change is that it is often difficult to understand exactly what is changing. In recent years, new technology has given us the ability to transform basic aspects of our lives: the way we converse and learn; the way we work, play, and shop; even the way we participate in political and social life.

Dissidents around the world use the Internet to evade censorship and get their message out. Cyber-gossips send dispatches to thousands via email. Musicians bypass record companies and put their songs on the world wide web for fans to download directly. "Day traders" roil the stock market, buying securities online with the click of a mouse and then selling minutes later when the price jumps.

This book argues that there is a common thread underlying such developments. It is not just a change in how we compute or communicate. Rather, it is a potentially radical shift in who is in *control*—of information, experience, and resources.

Part One, Revolution, explains how new technology is allowing individuals to take power from large institutions such as government, corporations, and the media. To an unprecedented degree, we can decide what news and entertainment we're exposed to, whom we socialize with, how we earn, and even how goods are distributed and political outcomes are reached. The potential for personal growth and social progress seems limitless.

This shift in control, however, is no sure thing.

Part Two, Resistance, shows how powerful entities are trying to limit our new digitally enabled autonomy. Some governments, for example, are restricting our access to certain content and preventing us from taking advantage of certain technologies. Some corporations are manipulating our information choices while creating the illusion of personal freedom. And we, unwittingly, may be their accomplices. Seduced by the rhetoric of individual power, we may not even realize that the old guard is still in charge.

At the same time, the new personal control is threatened by an equally menacing but less predictable foe: its own unyielding momentum.

Part Three, Oversteer, warns that individual control can be pushed too far. Enthralled with the idea of taking power from politicians, media giants, and price-inflating middlemen, we may lose sight of the benefits of representative democracy and of the need for intermediaries who bring us reliable news and high-quality products and services. Comforted by the sanctuary of filtered order in a world of sensory overload, we might unintentionally narrow our horizons, depriving ourselves of opportunities. Cherished values like community, free speech, and privacy could be diminished.

Part Four, Balance, charts a path between this Scylla and Charibdis. It describes how we can reap the benefits of the new control without succumbing either to resistance or to excess. To preserve democracy, truth, and individual well-being in this uncertain age will require a renewed sense of personal responsibility and commitment to our communities, as well as a fresh approach to governance that takes into account the shifting of control from institutions to individuals. We must achieve a balance of power for the digital age—between self-interest and public interest, the market and government, personal control and shared power.

That is an extremely abridged summary of my argument. What it doesn't explain is the host of interesting controversies and policy issues through which we will trace the new regime of changing and contested control. This cornucopia includes: cyberporn and censorship, customized news delivery, electronic commerce, online democracy, Microsoft's market power, encryption and law enforcement, copyright in the digital age, virtual communities, Matt Drudge, privacy, and the role of interactive technology in struggles against political tyranny.

My goal, it should be clear, is not to treat each of these issues comprehensively, but to use them to illustrate both the ways in which new technology allows power relations to be transformed and the ways in which different actors are responding to this possible sea change.

The reader should bear in mind that the viewpoints of the first three parts of the book are meant to be quite distinct: Part One presents the control revolution in the most optimistic light (with Chapters 4 to 6 particularly geared toward those who are not that familiar with the Internet), whereas Parts Two and Three take much less sanguine views. Part Four attempts to harmonize these competing voices, and a preview of my conclusions there may be helpful. There are six ideas that I believe should guide us as we respond to new technology's impact on society. Each is a call for balance between competing values:

1. *Rules and contexts*: When technologies change, it may seem that old rules no longer work. Some observers claim, for example, that digital technology makes copyright law obsolete, while the FBI says advances in encryption are undermining conventional law enforcement. Yet generally, the principles that underlie existing rules should still apply. We just need to map those principles in a way that respects new contexts.

2. *Convenience and choice*: Antitrust regulators claim that Microsoft's software design and business practices inhibit competition, while Microsoft responds that it is just trying to make computers and the Internet easier to use. Whatever the law says, the rub is that both claims are right. Consumers must be mindful of the trade-offs between convenience and robust choice.

3. *Power and delegation*: The Internet gives individuals the ability to bypass many intermediaries—in commerce, culture, and politics—and thus to make decisions that traditionally were made for them. Empowering as this may seem, real personal authority is about knowing when to make choices for yourself and when to let others whom you trust make them for you.

4. *Order and chaos*: New technology gives people the ability to personalize the information they receive and the social environments they inhabit. Yet as we order our worlds according to our own de-

sires, we must not forget the value of remaining open to serendipitous encounters.

5. *Individual and community*: The global reach of the Internet gives individuals an unparalleled degree of access to people, resources, and experience. All this potential will amount to little, though, if people use technology to ignore their local communities and commitments. We must use the Internet both to explore globally and to engage locally.

6. *Markets and government*: With the help of technology, individuals should increasingly be able to use market forces to their advantage. But relying excessively on markets threatens both equity and efficiency. Government has important roles to play in solving social problems, maintaining fairness, and protecting democratic values.

Finally, a word about perspective. Upon launching a new technology section in 1998, the *New York Times* ran advertisements that asked readers: Are you a technophile or a technophobe? My answer to that question would be neither, or perhaps both.

Technology is not like anchovies, which some people love and others hate, nor is it like the right to an abortion, which some are for and others are against. Rather, it is an indelible feature of our cultural environment—one that we must strive to understand, in all its gray-shaded complexity, so that we can make it as consistent as possible with our personal and collective values. Toward that end, my goal here is to present a forward-looking yet unvarnished view—what I would call a technorealist view—of how new tools such as the Internet are changing our lives.[1]

Part One

Revolution

"I would have nobody to control me; I would be absolute . . . ."

Cervantes, Don Quixote[1]

Chapter 1

"We Have Revolution Now"

O n August 19, 1991, I received a remarkable fax from a friend in Moscow. That morning, news alerts had informed the world that a coup had been launched by communist hard-liners against the reform government of President Mikhail Gorbachev, but media coverage was spotty and little else was known. Even President Bush and his staff were in the dark.

"We really aren't sure what's happening at this point," said one U.S. official. "We have no independent knowledge of what's going on," said another.[2]

Phone lines between the USSR and the U.S. were reportedly down. But somehow my friend Oleg, whom I had met two years earlier when I visited the Soviet Union as a student, managed to get a message through to me in New York using the fax machine in the office where he worked as a translator. In frenzied desperation, he wrote:

> I don't know how long it will be possible for me to use this channel of sending information. Situation is changed every minute. Two hours ago all Soviet Radio and TV programs began to read official propaganda messages of State Committee of Extraordinary Situation, which was organized last night. . . . So, it is military coup d'etat.
>
> Some independent Moscow broadcasters were stopped working this morning. Very popular independent broadcasting "Echo of Moscow" was turned off at 7:55 A.M. after they told that tanks are

near Moscow. You can't imagine what I do feel now. I am afraid of civil war. . . . If you need any information from me send me fax A.S.A.P. This is only thing that I can do now for my poor country.

I was working as an intern at a national weekly magazine at the time. We were, of course, hungry for information about the putsch. I quickly scribbled a reply, asking my friend a number of questions. Three hours later, there was another fax from Oleg:

There are a lot of tanks in the city. I counted more than 40 in my district. They are going to Kremlin. Downtown is full of tanks and soldiers. I have seen more than three battalions near the American embassy. There are many guns and cannons in the center of Moscow.

People here are shocked. . . . waiting for information about what's happened. All broadcasting and TV programs are still transmitting propaganda documents I wrote about before. . . . All mass media are under control now. It's terrible.

In a handwritten scrawl at the bottom of the fax was a postscript: "The clouds are gathering over the city. It'll be storm."

Over the next few days, a truly historic battle for freedom was waged in what we now call the former Soviet Union. But curiously, there were no more faxes from my friend. Only on August 26, a week after his initial message, were we able to reestablish contact. By that time, the world had learned that the pro-democracy forces had successfully put down the coup. A fax from Oleg explained his silence. Just hours after he had sent his second message, his international phone service had been cut off:

"What could I do? So I spent two nights and one day near our 'White House.' It was like a dream, but Kafka dream. . . . In any case, today I live in free country. A lot of things are changed or going to be changed very soon. We have revolution now."

▶

We have revolution now. Oleg was referring literally to the transition from Soviet communism to free-market democracy that was sweeping Eastern Europe—a series of events that signaled the end of the cold war and the rise of global capitalism. But there was another remarkable shift implied by the *way* in which he and I were communicating. Back in 1991, fax machines were still new enough that Oleg's ability to transmit instant, detailed reports of the Soviet Union's demise halfway across the world seemed itself to be revolutionary.

The newly minted independent media in the USSR had been neutralized by the old guard. In the West, journalists and heads of state alike were groping for the facts on the ground. Meanwhile, my humble twenty-one-year-old pal was able, for a time, to bypass official channels of communication to provide a firsthand account of a nation in tumultuous transition.

Oleg's ability to get information out of his country was actually only half the story. The revolution in Eastern Europe that took place between 1989 and 1991 owed much to the fact that new technology allowed dissidents to receive information—information that the ruling elites did not want them to have. This was a classic chapter in the ongoing historical relationship of technology to knowledge, and knowledge to power.

The highlights of this epic are familiar. In the wake of Gutenberg's fifteenth-century printing press, the availability of noncanonical religious tracts, most notably Luther's Ninety-five Theses, challenged and ultimately undermined the authority of the Roman Church. As printed works became available, individuals could for the first time begin to exercise real discretion over their information intake and their beliefs. As a sixteenth-century historian described it, "Each man became eager for knowledge, not without feeling a sense of amazement at his former blindness."[3]

As printing methods improved in the seventeenth and eighteenth centuries, books circulated more widely, and literacy and education blossomed. The scientific advances of Copernicus, Galileo, and Newton became widely known. The Enlightenment philosophies of writers like Locke, Rousseau, and Paine found an audience—Paine's "Common Sense," for example, sold 120,000 copies in three months[4]—

and ultimately popular expression in the rise of the republic in Europe and America.

In the nineteenth and twentieth centuries, the advent of mass media—the rotary press, penny papers, photography, film, radio, television—helped to pull together increasingly large communities and to spur cohesive modern nation-states. In certain parts of the world, like the United States, constitutional safeguards led to the emergence of a vibrant free press. In areas like the Soviet bloc, however, mass media mostly meant an information monopoly for state propagandists. Industrious folks behind the iron curtain might have managed to procure some clandestine samizdat materials. But with their gulags and secret police, the autocrats still had the upper hand in the battle over the free flow of information.

During the late cold war years, though, as satellite, video, and microprocessors proliferated, it became increasingly difficult for dictators of the world to regulate knowledge. Oleg and I were a case in point. From the time we met in 1989, I had been faxing him clips from the Western press and peppering my messages with the latest news about uprisings in the communist satellite states. And this kind of information was flowing in to Oleg and others like him via other electronic media, as well. In the Baltics, television broadcasts from Northern Europe, including reruns of American programs like *Dallas* and *Dynasty*, seeped across the border. Video- and audiotapes with dissident messages circulated throughout Eastern Europe's underground. (In fact, when I visited the USSR in 1989, the most valued Western commodities were not Marlboro cigarettes or blue jeans, as conventional wisdom had it, but blank cassettes.) And in universities throughout the region, computer users were starting to keep in touch with colleagues around the globe using a growing computer network that would come to be known as the Internet.

The same uninhibited exchange was going on around the world. In China, the students of Tiananmen Square faxed pleas for help to the West and received faxes from supporters abroad giving them vital information and encouragement.[5] In Central America, activists used shortwave radio to communicate with allies in the U.S. In short, technology was gradually giving individuals everywhere the ability to take control of information that was once parceled out exclusively by the state. Corrupt officials might have succeeded, for a time, in exercising their military might, but their efforts to hide the truth from their own people or the outside world were becoming futile.

"A Whole New Technology"

I was reminded of this new reality a little more than five years after Oleg's faxes when I sat in front of my home computer listening to programming from a Belgrade radio station that was being "broadcast" over the Internet.[6] It was December 1996. The war in the former Yugoslavia had ebbed and Serbian democracy activists had won local elections, but the authoritarian ruler Slobodan Milosevic had nullified the vote. Anti-Milosevic protesters filled the streets of Serbia's main cities and were emboldened by an independent radio station, Radio B92, which regularly aired updates about the protests.

Recognizing the station's power, Milosevic forcibly shut it down on December 3. Since he already controlled Serbia's TV stations, Milosevic must have thought this would fully silence the opposition. But in no time, Radio B92 rerouted its programming to the Internet where it was available, in digitized form, to computer users in Serbia and around the world. News of the activists' feat immediately flooded the email boxes of government officials, humanitarian groups, journalists, and supporters.[7] So successful were the Belgrade activists at getting their message out over the global network that, within two days, they had garnered enough international support to force Milosevic to let them back on the air.[8]

"The irony is that the Government meant to silence us, but instead forced us to build on a whole new technology to stay alive," said Drazen Pantic of Radio B92. "The drive to close us down has given us a tool to vastly expand our audience."[9] (The power of that tool was evident again in spring 1999 when Milosevic shut down the station to defy NATO, only to find it drawing a global audience on the Internet.)

The Radio B92 cybercasts continued the innovative political use of new media that had begun in Eastern Europe and China, but also represented an important advance. Though the Internet existed in the late 1980s and early 1990s, it was not yet widely used. The Serbian democrats were among the first activists to utilize the network in a way that tangibly and immediately affected global politics. By doing so, they amply demonstrated its unique strengths.

The Radio B92 activists showed, for example, how the Internet allows an individual to send a message—be it text, audio, video, or some combination—to hundreds or thousands of destinations as easily as to one, with no discernible increase in cost or time. Thus they could instantly and cheaply get their programming out to listeners around the world, along with text and pictures of the Belgrade street protests.

To experienced Internet users, this may seem old hat—but not when compared to other communications tools, even those that we consider fairly new.

Had Oleg been able to send his alerts to me by email, for example, the Soviet hard-liners probably would not have succeeded in silencing him. When they shut down international phone lines, they made his fax machine useless. But with the Internet he could have emailed his bulletins to an endless number of individuals in the USSR with confidence that someone would have had the means to forward them out of the country—by satellite or other advanced wireless technology. He could have been in immediate contact with an endless number of citizens around the Soviet Union and abroad, as the Belgrade activists were when they coordinated protests and international support in cities in Serbia and around the world.[10] And, like them, he could have sent sound and images along with his written updates.

Just a few years after the popularization of the Internet, the Radio B92 incident made it clear that this simple combination of computers and modems combines some of the best elements of all preexisting media. Organizers can, for example, merge the pinpoint accuracy of one-to-one technologies like the telephone with the broad reach of expensive media like television. They can use the Internet to create both a permanent archive, covering every detail of their struggle, and an instant alert network that will inform the world within seconds of late-breaking developments. Most important, they can create vibrant streams of conversation that are as free from restraint as any we know—

streams that can, within hours, grow from a rivulet into an unstoppable rapid.

This is one of the true marvels of interactive technology: the instant ability to spread your unexpurgated words—a piece of yourself, really—to the four corners of the earth. Even in the rush of millennial tidings, the singularity of this achievement cannot be overlooked. It is a privilege that would stir envy in the hearts of history's most powerful rulers and statesmen, not to mention fear. The Serbian democracy activists must have realized this, because they began to refer to their struggle against Milosevic as "the Internet revolution."[II]

A Fragile Reordering

If the phrase "Internet revolution" sounds familiar, that's because technology talk these days is suffused with references to revolution. Of course, we don't generally mean to conjure up images of tanks in the street or Che Guevara types. Rather, we speak of a communications revolution, an information revolution, a digital revolution. And the rebels in our midst are CEOs of huge telecommunications companies boldly "synergizing" their way into multimedia or young software entrepreneurs hoping to get lucky with "the next big thing." There is something undeniably convenient about these allusions to revolution. They are shorthand for a presumed common understanding of all that is happening because of the creeping ubiquity of new technology.

> Terms like communications revolution and information revolution don't go far enough.

But as linguistic proxies, they are also a bit perplexing. What do they mean in relation to the other things we have called revolutionary—say, the American or French Revolution, the scientific revolution or the industrial revolution? Or, for that matter, the upheaval in Eastern Europe a mere decade ago?

Each of those revolutions signified a distinct break with the past, the rise of a new order. Is that what we are experiencing because of the Internet and other new media? Certainly, we are communicating in ways that are different from before. And we have a whole new way to access and manipulate information. These are, no doubt, major developments. But in an age of unchecked hyperbole, it makes sense to ask: Are these changes really *revolutionary*? And if so, exactly what type of revolution are we experiencing?

To make sense of these questions, we need to probe deeper: How will the Internet affect our social lives, our jobs, and our perspective on the world? What will it do to our basic relationships with family and friends, with neighbors and far-flung fellow citizens, and with the powers that be in government and the corporate world? Will it enhance or diminish core democratic values like freedom, equality, community, and social responsibility?

Refocusing the inquiry this way requires us, I believe, to look past the terms we commonly bandy about to find a new description that captures more accurately what it is that is changing. Terms like *communications* revolution and *information* revolution actually don't go far enough in explaining the transition at hand. There is more at stake here than being able to send messages more quickly or having access to a supercharged digital library. That's just the immediate effect. It's like when a stone is thrown into a pond: the splash catches your eye first, but it's the endless ripples that have the broadest impact and are most interesting to observe.

In this case, there is a pattern to those undulations. What they suggest is a potentially momentous transfer of power from large institutions to individuals. The real change set in motion by the Internet may, in fact, be a control revolution, a vast transformation in who governs information, experience, and resources. Increasingly, it seems that *we* will.

To be sure, "individuals" and "institutions" are not monolithic entities.[12] But these terms do capture something fundamental: the palpable sense of deciding for yourself as opposed to having some larger, impersonal *them* deciding for you. This includes choices about intake of news and other information, social interactions, education and work, political life and collective resources. It is a time of diminishing stature for many authority figures: legislators and other public officials, news professionals, commercial middlemen, educators. Hierarchies are coming undone. Gatekeepers are being bypassed. Power is devolving down to "end users."

The upshot of new technology, then, seems to be its ability to put individuals in charge. Yet what makes this upheaval so much more authentic than those revolutions described by Panglossian futurists is its volatility and lack of preordained outcome. Contrary to the claims of cyber-romantics, individual empowerment via technology is not inevitable.[13] Rather, it faces predictable and unpredictable challenges. It will likely be defined by protracted struggle, a clash of values, and a fragile reordering of the social landscape that could come undone at any time. That is why the word "revolution" actually describes well the shift in control made possible by the Internet.[14] Some institutional forces are resisting, and will continue to resist, giving up power to individuals. And there is a danger that some individuals will wield their new control carelessly, denying themselves and others its benefits.

Still, the resemblance to political revolution is, in important ways, only metaphorical. Computer nerds aside, there is no junta driving this process of change. In a sense, we are all its protagonists (whether or not we know it). The control revolution, in fact, has some of the texture of a subtle historical shift such as the agricultural revolution or the industrial revolution—not in the sense that it will be centuries in the making, but because it may emerge undetected. At the same time, we will see that institutional resistance to this change may be just as inconspicuous.

Not every form of individual empowerment by technology, though, is a zero-sum game. In addition to assuming command of functions once managed by others, individuals are increasingly able to control aspects of life that previously were outside anyone's dominion. New media, we will see, allow us to manipulate and even conquer some of the limits of time and space.[15] And outside the realm of communications, other innovations also are presenting us with remarkable new opportunities to shape our world.

"Individuals are acquiring more control over their lives, their minds and their bodies, even their genes," says *New York Times* writer John Tierney.[16] Biotechnology allows us to know things about our physical and psychological makeup that once were unknowable, such as the likelihood of getting cancer or of having a predisposition toward violence. New developments in science increasingly will let us control our health and well-being, and our natural environment, in ways that we never could have before. From genetic screening to "the wholesale alteration of the human species and the birth of a commercially driven eugenics civilization," as author Jeremy Rifkin describes it, technological advances will allow individuals to make unprecedented decisions about their lives (and the lives of their offspring), decisions that will raise thorny issues of ethics, spirituality, and fairness.[17]

Important as these and other technological developments are, I will focus in this book almost exclusively on communications technologies. Still, there is a lesson to learn from the coming biotech battles as we consider the control shift made possible by new media such as the Internet: No one doubts that decisions made by governments and corporations on issues of cloning, gene therapy, and the like, are political and highly charged. Similarly, few would deny that even individual decisions about biotech may be matters of both personal and public concern. Yet for some reason our assumption is the opposite when it comes to the design and development of communications tools, and the manner and environment in which we utilize them. We tend to see these as apolitical choices of significance only to narrow constituencies.

The truth, though, is that these decisions about communications technology may affect who we are socially and politically as much as biotechnology can alter who we are genetically and physically. To begin to understand this, we first have to become more familiar with the features of new communications technology. Only then will we start to appreciate what these features make possible and why they are at the heart of an unfolding battle for control.

Chapter 2

The Politics
of Code

"Guns don't kill people. People kill people." This slogan of the National Rifle Association may seem like a strange place to start in trying to understand the impact of the Internet. Yet when we look at how fairly uncomplicated tools such as guns affect society, we can see aspects of technology's impact that may not be as evident in a more complex setting such as computing.

The NRA's motto is meant to show how foolish gun-control opponents are for holding firearms "responsible" for violence, instead of focusing on those who use guns unlawfully. And in a literal sense, the NRA slogan is right. Guns don't kill people on their own (and neither do bullets). To suggest that they do is to ignore the importance of how technologies are used. Indeed, the way we use guns would seem to be the single most important factor in determining their impact on society.

But is it? Handguns do not grow on trees. Nor are they inevitable by-products of human civilization. Like all technologies, they are deliberate human creations embedded with values. In fact, they have features that might cause them to be predisposed more toward some outcomes than others. This is apparent when we consider whether handguns should be built with safety locks, which might greatly reduce the number of accidental shooting deaths.[1] The ultimate impact of handguns, then, has much to do with the way they are designed.

The NRA's slogan is a telling example of what we might call the myth of neutral technology, the idea that artifacts are not political because they can't do anything on their own.[2] It is a presumption whose folly is matched only by its opposite: the notion that political outcomes can be determined *solely* by a technology. A more fruitful way to think about technology's impact is that it depends on a combination of design, use, and the environment in which it is deployed.

Technologies *are* political, then, in the sense that they have proclivities that can be easily tapped by certain policies and practices. And the design of the technology itself may well cause it to lean slightly, or not so slightly, toward one outcome rather than another. (Think of a bomb and an artificial heart. Whatever its potential for deterrence, the bomb facilitates a negative outcome more easily than a positive one, and the opposite is true for the artificial heart.)

Recognizing the contingent nature of technology design is particularly important when it comes to communications technology because of the rapid evolution and convergence of different media. Most significantly, video, voice, and data networks—which arrive in our homes today via TV, phones, and computers, respectively—are becoming integrated. With its abrupt growth, in fact, it sometimes seems that the Internet will swallow up all traditional media. Yet the Internet itself is becoming hard to define. Already you can surf the web via the coaxial cable that brings you cable TV and get email on a wireless portable phone. (For convenience's sake, I'll refer to this bundle of communications tools—the Internet, interactive television, digital communications appliances, and their twenty-first century incarnations—as the Net.[3])

One thing we can say for sure about the Net is this: Just as it is growing quickly, its form can also be changed quickly. That's because although the Net depends on physical hardware—networks of computers and wires—it is defined mostly by *code*. Code is the stuff that computer programmers create: software, technical protocols, network designs. It determines how information flows online and who can control it. Yet code is not immutable. In fact, code can usually be altered quite easily—sometimes with a few keystrokes.

Contrary to the common pronouncements that the Internet is "inherently democratic,"[4] then, this technology is not automatically a guarantor of liberty or fairness. Indeed, as Harvard law professor

Lawrence Lessig and others have persuasively argued, code may be at the heart of various power struggles in the digital age.⁵ In fact, as I will discuss in Part Two, some institutions may respond to the new individual control by trying to shape the code of the Net to retain their authority.

The Code of the Net

In light of the contested nature of code, it is important to identify the proclivities of different code features (while recalling that it takes human agency to set those potentials in motion). Below are six code features of the Net that can enhance individual control. Four of these features characterize today's Internet, though they cannot be taken for granted. Two others must be achieved in order for the Net to meet its fullest potential for individuals.

Starting with the existing code features: the Net is, first, characterized by *many-to-many interactivity*. Complicated as this may sound, there's nothing unusual about interactivity. After all, the telegraph and telephone are interactive. But they only allow one-to-one communication between two parties.⁶ Mass media such as television and newspapers, on the other hand, are one-to-many but they're not interactive: One broadcaster or publisher can speak to a huge audience, but those who watch or read can't easily speak back. With the Net, however, communications can be one-to-one (email, for example) or one-to-many (when one puts up a web site or sends a message to an email list). This means that, in the big picture, the Net is many-to-many, because "many" people (in fact, anyone with access) can speak to "many" others.

Many-to-many interactivity has rightly been hailed as one of the most potentially democratic aspects of the Net because it allows individuals to be creators of content rather than just passive recipients, and active participants in dialogue instead of just bystanders. It is how a single Radio B92 activist can reach out to the whole world and get instant replies from anyone she contacts. Not surprisingly, this has made the Net's interactivity a prime target of those who are anxious about the control revolution.

Second, content on the Net is *digital*, which means most importantly that it is flexible—in terms of how it can be stored, used, and manipulated. In analog format, information is wavelike and imprecise. In digital format, information is represented in binary fashion as one of two numbers, zero or one. Because of this simplicity, digital data can be perfectly replicated with little effort. Unlike analog information, the clarity of the message isn't lost from copy to copy, or over long stretches of travel. The information can also be compressed and manipulated easily. Digital signals, whether they represent audio, video, or words and numbers, can be merged and carried together—for example, on one compact disk. All this adds up to tremendous flexibility and power for the user. This is the primary reason that formerly analog media—like television, the telephone, and even stereo systems—have gone digital. The flexibility of digital information is crucial to the rise of the new individual control.

Third, the Net is a *distributed, packet-based* network. As television networks and airline routes demonstrate, a network can have one origin or it can have a few major hubs. It can also, however, be more fully decentralized so that information travels from place to place without having to return to any central point. The fact that distributed networks are not oriented toward centers is significant in terms of individual control, because it means that end users have a greater ability to dictate the flow of information. Along with interactivity, the distributed nature of the Net helps individuals to bypass gatekeepers. Related to this is the fact that the Internet is a packet-based network, rather than a circuit-switched network like the telephone system. This means that messages are broken up into packets; they each carry "directions" that allow them to travel separately to a destination and then join together again. This increases the network's efficiency and, significantly, the ability of users to route around censorship.

Fourth, the Net is *interoperable*, which is basically a technical way of saying that it is open to all comers. Interoperability means that hardware and software are designed so that information can flow freely throughout a network without bottlenecks or barriers. Without this principle, the Net's central feature, its "networkness," is undermined. Instead of getting access to the resources of their choice and communicating with anyone anywhere, users remain stranded on remote data islands, able to communicate only with their fellow captives (that is, others who use the same closed system). An important feature of the Net's openness is the fact that its basic protocol of information exchange, TCP/IP (Transfer Control Protocol/Internet Protocol), is nonproprietary. Like a set of grammatical rules, TCP/IP is not owned by anyone.

In terms of efficiency, an open, interoperable network is generally preferable to a series of smaller, exclusive networks. For democratic purposes, a single network is clearly better because anyone can communicate with anyone else. (It's the same reason we have one phone network, instead of the multiple nonconnecting systems of the early days of telephony: The more people who can be reached, the more value the network has for each individual user.) Notwithstanding the misguided efforts of some companies to erect blockades on the Net, interoperability should continue to be a prominent code feature.

Two other code features need to be achieved: First, the Net should be *broadband*, which means that it must have a large bandwidth or carrying capacity. Broadband networks allow users to easily get access to complex information such as video images. Coaxial cable and fiberoptic lines are broadband, but the old phone lines that still carry most Internet traffic into households today are not. Experience shows that not every aspect of tomorrow's Net needs to be broadband in order for it to be effective, but the greater the bandwidth the more efficient the network. A variety of information—text, audio, video—can be transmitted via a single pipe. Fortunately, there is good reason to believe that broadband networks, which are now in their infancy, will soon be standard.

Second, access to the Net must be *universal*. This is not strictly a code feature because it is more socioeconomic than technical. But it is important enough to the control revolution that it deserves to be seen as an extension of the technology. Universal access requires that all individuals have some meaningful access to the Net. Today we are far from that ideal, though important progress has been made, at least in developed nations, since the time when the Internet was used exclusively by elites in government, the academy, and industry. (The percentage of Americans who use the Internet grew from 14 percent in 1995 to 41 percent in late 1998.[7]) Points of access might be privately owned computers in the home or at work, or public terminals in schools, libraries, and community centers. As with interoperability, there are political (and economic) advantages to making sure everyone has access to the network—which will, in the twenty-first century, be the gateway to informed citizenship.

Whether the expansive Net of tomorrow will have all six of these features is difficult to say. It is notoriously difficult to know how technologies will evolve. Thomas Edison, after all, originally conceived of his phonograph as a way to record one's thoughts and send them to friends. And the telephone was used early on in a few places to pipe music into homes from distant concert halls. Trying to assess the long-term social and political impact of a technology is even trickier. Certainly no one predicted that the television would become a kind of baby-sitter.[8] The rapid pace of the Net's growth makes it even more difficult to assess its impact.[9]

Still, by focusing on these basic code features of the Net, we will at least be aware of how the Net's potential will evolve and change. At the same time we cannot forget to consider how the Net is used. And, as noted above, there is yet another feature to take into account: the broader social environment in which the tool exists. The ultimate effect of technology is a factor not only of design and use, but of prevailing mores and ideologies.

The Exaltation of the Market

In the case of the control revolution, the significance of the surrounding environment is abundantly apparent. This shift in control is reinforced by—and it reinforces—current trends in political and social thought that emphasize the power of the individual.

One of the most prominent of these trends is the exaltation of the free market that has occurred in recent decades. Public, collective control of resources has given way to private, individualized control. Laissez-faire economic policies have become dominant in the major nations of the world because of the belief that the economy works best when actors in the private sector control the production and distribution of goods and services. Public policy is driven less by abstract principles of social justice and the common good than by a desire to achieve efficiency and to satisfy the "revealed preferences" of consumers. Autonomy is favored; restraints on choice are frowned upon. Indeed, notions of market freedom and personal freedom have become intertwined in public discourse (and sometimes confused, as well).

The dominance of the market is notably evident in the way societies manage their communications resources. Telephone networks, radio and television stations, and newspapers are privately owned in the U.S. and the Western world. Though some electronic media are regulated in the public interest, scholars and policymakers increasingly believe that there is no longer justification for state intervention in the information marketplace. Once the prevailing notion was that broadcast regulation was necessary because of spectrum scarcity: the physical limits on the number of frequencies available for radio and television broadcasters. Those who received coveted broadcast licenses were therefore obliged to serve the public interest. And regulation of telecommunications was justified by the fact that phone networks were perceived to be a natural monopoly, a resource so essential that it made sense to have one carrier with rates and service requirements regulated by the state.

Now, though, the conventional wisdom has shifted. The emergence of new technologies with seemingly unlimited channel capacity has caused scarcity to be seen as a moot issue. Technological advances, moreover, have led most observers to believe that competition is possible in markets that once were regulated monopolies.

The Telecommunications Act of 1996 represented a major step by the federal government in this deregulatory direction—an endorsement of the idea that broadcast and telephone networks, like print, should generally not be regulated. And on July 1, 1997, just days after the Supreme Court struck down the Communications Decency Act as an impermissible regulation of free speech online, President Clinton announced a similar hands-off policy for cyberspace. Governments, he said, should not "stand in the way" of the Internet, but should simply enforce "a predictable, minimalist, consistent and simple legal environment for commerce."[10] The U.S. would refrain from taxing or otherwise interfering with the Internet, treating it like "a global free-trade zone."[11]

One irony of this approach is that the Internet exists today only because of the foresight and largesse of the federal government. In the 1960s and 1970s, the Defense Department's Advanced Research Projects Agency created a computer network to allow defense researchers at scattered universities to share computing resources. (Contrary to conventional wisdom, the Internet's origins were primarily in facilitating research, not in allowing the U.S. communications system to survive a Soviet nuclear attack, though that was a recognized benefit of the decentralized network.[12]) Soon the researchers began using the network, called Arpanet, more to exchange email than to share data and computing power. Arpanet expanded to other academic and research uses and grew, under the auspices of the National Science Foundation, until it eventually mushroomed into the network of computer networks that we today called the Internet.

In the mid-1990s, as the Internet became more commercial, the federal government began to withdraw from funding and overseeing it. In 1995, it sold off the network backbone to a private consortium of large technology corporations, and it gave one company the exclusive right to register domain names like nytimes.com and aclu.org. The Internet was essentially privatized. Today, while the federal government still plays a role in funding research and development in this area, its day-to-day oversight of the Internet is almost nil.

> The control revolution represents the merger of the communications revolution and the free-market revolution.

This laissez-faire approach to communications technology occurs at a time of increasing reliance on market-based solutions to a wide variety of social problems. Privatization has even come to sectors like health care, education, and social welfare, on the presumption that these essential services can be provided most efficiently when decisions are made by individuals and entities in the private sector, rather than by government. It's a presumption that is, at best, untested; at worst, it could have corrosive effects on public welfare and social justice.

Yet whatever one thinks of the increasing marketization of society, it is difficult to deny that mainstream politics and economics today rely increasingly on an abstract faith in individual control, as opposed to collective control. In this sense, the control revolution represents the merger of the communications revolution and the free-market revolution.

The Culture of Individualism

In the United States, personal power is buttressed also by a strong civic tradition of individualism. Today that tradition may be most evident in various activities that are meant to help us to take command of our lives: self-help and recovery, spiritual awareness, diet and exercise. But there are strong historical roots undergirding this focus on the self.

From the early colonial spirit of exploration, independence, and frontier conquering, to the emphasis in our Constitution on individual liberty (as opposed to community), to the nineteenth-century transcendentalism of Emerson and Thoreau, to the soul-searching of the "me" generation of the 1960s and 1970s, it's something of a cliché that Americans are a highly individualistic people. As Tocqueville pointed out, Americans are self-reliant, skeptical of authority, and inclined to "imagine that their whole destiny is in their hands."[13] It is thus not surprising that we would develop and come to rely upon tools that allow us to broaden our swath of personal dominion.

This emphasis on the individual has also been matched recently by a steady erosion of trust in various institutions. From the 1960s to the 1990s, Americans' confidence in government has fallen precipitously. Three-quarters of respondents in 1964 said they trusted the federal government to do the right thing most of the time. Now only a quarter of Americans say this. Other American institutions have suffered a similar loss in stature over the last three decades. Public confidence in universities has gone from 61 percent to 30 percent; in major companies, from 55 percent to 21 percent; in medicine, from 73 percent to 29 percent; in journalism, from 29 percent to 14 percent.[14]

Similar trends are evident in countries such as Canada, Britain, Italy, Spain, Belgium, the Netherlands, Norway, Sweden, and Ireland.[15] Perhaps challenging the long-standing belief that American individualism is unique, popular European authors now write of "the new era of personal sovereignty" and the age of "the sovereign individual."[16]

Indeed, the desire for personal control may be consistent with the idea of freedom in all of Western post-Enlightenment thought. The philosopher Isaiah Berlin, in his landmark definition of positive liberty, says that it "derives from the wish on the part of the individual to be his own master." As he explains:

> I wish my life and decisions to depend on myself, not on external forces of whatever kind. I wish to be the instrument of my own, not other men's, acts of will. I wish to be a subject, not an object. . . . I wish to be somebody, not nobody; a doer—deciding, not being decided for, self-directed and not acted upon by external nature or by other men as if I were a thing, or an animal, or a slave incapable of playing a human role, that is, of conceiving goals and policies of my own and realizing them.[17]

This is what the control revolution seems to be all about: following one's own will. Deciding, not being decided for. Conceiving goals and being able to realize them.

The Psychology of Control

Psychologically, the desire to obtain personal control is not surprising. Scholars from a variety of psychological schools of thought see the attempt to master one's environment and to attain increasingly higher degrees of control as dominant forces in life.[18] Alfred Adler, for example, saw this innate need as the source to which "all social behavior can ultimately be traced."[19] Stanley Renshon argues that "the need for personal control is . . . an existential 'given.'"[20] Control gives us satisfaction, he says, not only because it allows us to achieve gratifying outcomes, but because we are able to minimize anxiety that might be present if we were not in control.[21] In other words, our interest in personal control is motivated as much by a survival instinct as by narcissism. It is key to our sense of self-esteem and confidence.

How does this need for personal control play out in everyday life? Consider the fact that more people are afraid of flying than of driving, despite statistics showing that air travel is safer than auto travel. Why is this? Probably because when we drive, we're pretty much in control, whereas in a plane we surrender control to someone else completely.[22] Most of us therefore feel safer in a car than in a plane. This feeling may be irrational. Yet even the illusion of control "seems to be central to man's ability to survive and to enjoy life,"[23] in the words of one psychologist. Similarly, other researchers in this area argue that the "sense of personal control, whether valid or not, is prized in our culture."[24]

Of course, we don't need experts to tell us how important the feeling of personal control is. It's something each of us knows from our own development. We go from being dependent as infants to having some degree of independence as children to pushing the limits of that liberty as young adults and ultimately to a rough state of autonomy as mature adults.[25] Increasingly, we determine our daily lives, our decisions, our futures. But still we experience battles of control. We sense these tensions in our relationships with our loved ones—our spouses, parents, children, and friends—and with our acquaintances, employers, and colleagues. We face control issues in our habits of work, exercise, consumption, and spirituality. And control is a central factor in our political and economic lives, as we struggle to become stakeholders in society.

Questions of individual control are also essential to our relationship to nature and biology. The creation and use of technology is particularly central to our attempt to impose human will on circumstances and forces that once seemed ungovernable. From the invention of the stone ax to the advent of the steam engine, transistor, and jet airplane to today's genetic engineering, humans have always developed tools to achieve new forms of control—over resources and the material world,

over time and space, over our own physiology. Our toolmaking abilities have set us apart from the rest of nature's creatures and instilled in
us a persistent belief in the ability of artifacts to make us free, whether
that artifact is an abacus, a cotton gin, or—as the next chapter suggests—a computer.

Chapter 3

Gaining Control

On the first page of his bestseller *The Road Ahead*, Bill Gates writes that he was initially drawn to computers as a child because "here was an enormous, expensive, grown-up machine and we, the kids, could control it." Though the young Gates and his friends were too young to drive or do other adult things, in the realm of computers they could be kings. "We could give this big machine orders and it would always obey," he recalls. "To this day it thrills me to know that if I can get the program right it will always work perfectly, every time, just the way I told it to."[1]

Gates is not alone in feeling that computers put him in charge. In Steven Levy's *Hackers*, a young computer wizard points out that "the computer . . . was just some dumb beast following orders, doing what you told it to in exactly the order you determined. You could control it. You could be God."[2]

Statements like these seem to celebrate the ability to control the machine itself, to domesticate it like some wild beast. Yet claims of control increasingly go further. Sherry Turkle, a psychologist and MIT professor, says that computer hobbyists often report that their machines provide them with a sense of mastery and order that is absent from other parts of their lives.[3] As one frequent user puts it, "I can't control the rest of the world but I can control my computer . . . ."[4]

> What does controlling a computer have to do with controlling one's life?

There is something peculiar about this juxtaposition. What does "controlling" a machine, even a powerful computer, have to do with "controlling" one's life? How can playing computer games, for example, possibly compensate for powerlessness in work or politics?[5] One form of authority is limited to the operation of a box of silicon and wires; the other has to do with wielding influence in the real world. On a crude level, perhaps assertive behavior in one sphere of life can offset helplessness in other spheres.[6] But there is something deeper here: a relationship between the use of computers and the desire to increase one's relative control of the external world. To understand this connection, and particularly its relation to the Net, we need to step back briefly to consider how our attitudes toward computers have evolved.

Taming the Machine

Before PCs became commonplace in the 1980s, most people thought of computers (if they thought of them at all) as mammoth objects with inscrutable vacuum tubes, blinking lights, and whirling magnetic tapes. This fearsome vision of computers—straight from sci-fi movies and TV shows—was abetted by the fact that these machines were scarcely seen in everyday life. They were, for the most part, operated by highly trained scientists in the exclusive enclaves of the academy, government, and large corporations. But with the remarkable improvement of the microprocessor, computers became increasingly small, affordable, and available. They entered our workplaces, schools, libraries, and homes.

Still, in their early years, PCs continued to provoke confusion and suspicion among the general public. Those who used computers regularly were stereotyped as geeks, misanthropes, even malevolent misfits. This view was born mostly out of ignorance. But it also reflected the fact that, reasonably or not, most people saw computers as mind-numbingly complex, boring, and perhaps even belittling. Recognizing this, the smartest entrepreneurs set out to build computers that would empower individuals rather than intimidate them.

The creative minds at Apple Computer understood the challenge better than anyone. In 1984, Apple released the Macintosh, the first commercial PC with a graphical user interface and a mouse. Instead of having to plow through dark fields of cathode-green text, remembering obscure commands, one could simply point and click through a soothing facade of simple, aesthetically pleasing icons: a desktop, menus, folders and files, dialog boxes, a trash can. Where earlier computer designs had seemed alienating, cold, and confusing, the Macintosh was humanizing, warm, and likable.

"User-friendly" was the term that evolved to describe this new computer interface, yet there was more to it than just amiability and simplicity. Apple consciously sold customers on the idea that the Macintosh was a computer for a new type of user—for "the rest of us," as the company put it. Nowhere was this strategy more evident than in their brilliant television advertisement introducing the Macintosh. It ran only once, in January 1984 during the Super Bowl, but caused such a stir that many who didn't see it undoubtedly heard about it. In the ad, a young woman liberates a horde of downtrodden info-age serfs from the gray, sterile tyranny of Big Brother—in this case, not the intrusive state of Orwell's *1984*, but Apple's powerful and lumbering rival, IBM.

Where IBM's PC was the computer of the establishment, Macintosh would be the tool of the creative, the young, the cutting-edge. The Mac appealed to those who felt threatened—not just by the complexity of computers, but by the prospect of becoming a cog in some routinized authoritarian scheme. Later, when Microsoft copied the Mac approach with its Windows operating system, the genius of the graphical interface would become universally appreciated. But it was Apple that first realized what we needed in order to become comfortable with computers. We needed more control.

In part, this meant being liberated from the drudgery of a text-only interface with the PC. As Steven Johnson explains, the graphical interface gave us the benefit of "direct manipulation," the ability to "get your hands dirty, move things around, make things happen."[7] Freed from having to learn arcane written commands, users could concentrate on what they wanted their computers to do *for* them rather than what their computer demanded *of* them.

Word-processing programs gave us new authority over the written word. Unconstrained by the permanence of ink and page, we could free-associate, type aimlessly, then cut, paste, and rearrange. We could spellcheck, highlight, and change the look of our documents a hundred times. Spreadsheets and personal finance programs gave us the ability to manipulate and keep track of numbers in new ways. Graphics programs allowed even the least artistically inclined to appear talented. And an endless surfeit of games indulged the imagination. In all these cases, software designers were writing code with the specific goal of enabling the individual. Every new release came with more options, more opportunity for customization.

The Macintosh and the now-ubiquitous Windows operating systems have extended this ethic to the point of allowing each user to personalize even the basic elements of the interface. On a Mac, for example, a user can name her hard drive, change the background colors on the screen, rearrange the desktop, and customize menus and commands. Personal computers have truly become personal. Apple emphasized this point with its "What's on your Powerbook?" ad campaign, which showed unlikely pairs of individuals and lists of the sundry items—addresses, screenplays, math equations—they stored on their portable Powerbook computers.[8] It's not just a box, Apple was telling us, it's an extension of you.

Even for those of us who use computers on a daily basis, these individual-oriented features may go largely unnoticed, if not unused. But they represent a broader philosophy of computing and technology generally—one that encourages each of us to feel as if the machine works, as Bill Gates put it, "just the way I told it to." Personal satisfaction has become perhaps the central value in our use of computers. And why not? This new emphasis on gratification and comfort has allowed us to transform what were sources of confusion into sources of power.

Now, with the reach and flexibility of the Internet, we can extend that power outward: from the tidy, inward focus of managing our files and applications to an increasingly externalized control—a remote control—of our interactions with the information and inhabitants of the whole world.

Just as you grew accustomed to choosing fonts and customizing your desktop, you'll now choose online communities and customize your news. You saved time by not having to rewrite a cover letter; now you'll post an evolving resume online and—if you're fortunate to have the right skills and experience—find work easily. That finance program helped you balance your checkbook. Tomorrow you'll save money online by having a software agent find the lowest price for a product you want. (Indeed, software agents, or bots, will likely carry out all sorts of tasks for you.)

The same way that PCs enabled us to take command of aspects of our personal lives, the Net will give each of us more of an opportunity to take command of our interactions with the world at large.

Reflect again, then, on the relationship between controlling a computer and controlling one's life. The Net means that the two could become increasingly intertwined. Sitting in front of a computer screen in your bedroom or office, it may be hard to see how a click here and there can really affect anything beyond the box, let alone your four walls. But connect that box to an outside line and things start to happen. As you gain access to newly personalized information, your perception of the world changes. Publish online and caucus with fellow travelers, and you can change other people's views. Sign a digital petition and influence the political process. Buy stocks—or groceries—online and affect the economy. Telecommute from home and redefine work, while reducing auto traffic and air pollution. These are the types of individual actions that are helping to produce a revolution in control.

The Disappearance of Cyberspace

With all this potential for personal control, it's noteworthy that cyberspace has so far mostly been described as "out of control." Average users often find it confusing and disorderly. Critics describe a chaotic medium overrun by porn, hate speech, and mindless flame wars. But oddly enough, this exaggerated sense of clutter and disarray actually proves just how much the Internet is defined by the potential for individual control.

I recognized this when I was telling a friend of mine about the premise of this book. He looked skeptical. "Personal control?" he said. "When I go online, I feel totally lost. There's nothing to guide you or tell you what to do."

Exactly. No one is in control—except you. And if you're fairly new to the Net, then it may well feel frenzied and unmanageable. You're not powerless because someone else is pulling the strings, though. You're just beginning to realize that the strings are there for you to pull yourself.

In time, as we become more familiar with online interaction and navigation, and as new software tools are developed, the Net will become more domesticated. Design will likely be simpler, information will be easier to find, and the whole experience will become more predictable and routine. Norms and rules will develop. Soon enough, naysayers will realize that cyberspace is no more dangerous or perplexing than physical space. Indeed, it will become increasingly difficult to distinguish the two, as cyberspace will simply become a way of looking at the world.

Many of the digital vanguard seem to think otherwise. They have urged us to see cyberspace as if it were elsewhere, a place with its own law and sovereignty. In 1996, for example, cyber-activist John Perry Barlow wrote a Declaration of the Independence of Cyberspace, which minced no words about the illegitimacy of governments exercising jurisdiction over the "space" in which people interact online:

> Governments of the Industrial World, you weary giants of flesh and steel, I come from Cyberspace, the new home of Mind. On behalf of the future, I ask you of the past to leave us alone. You are not welcome among us. You have no sovereignty where we gather. . . .
>
> I declare the global social space we are building to be naturally independent of the tyrannies you seek to impose on us. You have no moral right to rule us nor do you possess any methods of enforcement we have true reason to fear.
>
> Governments derive their just powers from the consent of the governed. You have neither solicited nor received ours. We did not invite you. You do not know us, nor do you know our world. Cyberspace does not lie within your borders. . . .[9]

Barlow, a former Grateful Dead lyricist and self-described "cognitive dissident," would probably be the first to admit that his pronouncements are equal parts theater and theory. But he is not the only visionary who has described cyberspace as a place where communities exist, altercations occur, and cultural practices congeal.[10]

Some legal scholars even have asserted that cyberspace should have its own law and legal institutions, and have questioned whether state-based governments should have jurisdiction over online activity. We should, they say, see cyberspace "as a distinct 'place' for purposes of legal analysis by recognizing a legally significant border between Cyberspace and the 'real world.'"[11]

> Cyberspace is too important to be thought of as elsewhere. Rather, it is right here.

This cannot be right. Though the sentiments are well-intentioned and understandable—some Internet users may feel they are somewhere else when they interact online, and there are real legal difficulties that arise because of transnational communications—it would be a mistake, conceptually and practically, to erect a barrier between online and offline activity. Cyberspace is not somewhere "out there," a world apart from flesh and blood, asphalt and trees. Our actions online have (need it even be said?) a real impact on the lives of other human beings. When a fraudulent securities offering on the Net causes novice investors to be bilked of their hard-earned money, for example, that's a "real world" injury.

In short, cyberspace is too important to be thought of as elsewhere. Rather, we should think of it as being right here. In fact, it is so close to us, so increasingly significant and indispensable, that it will eventually recede from the fore and even disappear. Disappear, that is, in the same sense that the wallpaper pattern in your bathroom eventually becomes so familiar that it fades away and escapes notice.

A Lens on Life

Fascination with cyberspace's exotic unfamiliarity is to be expected, for this is how we treat every new technology at its inception. As online interaction becomes less foreign and more a part of everyday existence, though, it makes sense to think of cyberspace not as a place or even a metaphorical space, but as a lens on life. The Net is an interface with which we can do almost anything: learn, work, socialize, transact, participate in politics. It allows us to control other things—the information we are exposed to, the people we socialize with, the resources of the physical world.

Thinking of the Net in terms of control even makes etymological sense. The word "cyberspace," made popular by science-fiction writer William Gibson in the mid–eighties, derives from *cybernetics*, which is the science of "control and communications theory."[12] Cybernetics, in turn, was coined half a century ago by a group of scientists led by Norbert Wiener, and was based on the ancient Greek word *kubernetes*, which meant "steersman" (as in the steering of a ship) or "governor." Cyberspace, then, can be thought of as an interface of personal control—a way that we steer reality or govern life.

Governing life? By tapping a computer keyboard? Yes, it may sound a bit far out. Most folks are just trying to get connected to America Online or hoping to figure out how to get rid of all the junk email that is clogging their email boxes. Yet this somewhat bewildered preoccupation with the Net's mechanics will pass as the technology becomes increasingly familiar.

In the early days of the telephone, people shouted into the receiver and conversation was stilted, yet now phone interaction is as natural for most of us as face-to-face contact. In the first years of radio, families gathered resolutely around the console at fixed hours each week to listen to programs. Today, the radio is a constant companion: it wakes us up, keeps us company in the car, and envelops us in the supermarket and at the office. (Indeed, few people listen to the radio any longer while *not* engaged in some other activity—driving, working, cleaning house.)

An even better comparison might be the adoption of alphabetic writing or spoken language. We don't think about letters as we write or grammar as we speak (unless we're learning a new language). Alphabets and language are our most taken-for-granted communication tools. They are so familiar that they just disappear. Similarly, just as the original Macintosh operating system worked because it was a fairly unobtrusive interface between the user and the resources of the personal computer—and a good interface wants nothing more than to be invisible[13]—the Net will increasingly be our inconspicuous interface with the world, another taken-for-granted way of understanding and filtering reality.[14]

More important than the ordering of the Net itself, then, is the way that it will let us reorder our lives. In other words, as the mystery of Internet communication fades, we will concentrate less on the computer and the network, and more on what the technologies allow us to do. *Allow*, though, is the operative word. Just because technology enables a revolution in control does not mean that it's a sure thing. To start with, we have to make it happen.

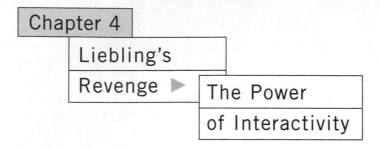

Chapter 4

Liebling's
Revenge ▶

The Power
of Interactivity

When the history is written of the clash between old and new media at the end of the twentieth century, journalist Philip Elmer-DeWitt should score a pivotal role. A writer and editor for *Time* magazine, Elmer-DeWitt has penned a number of important articles on technology and the Internet. But the story he'll likely be remembered for is one that he would probably rather forget. It was a special feature about Internet pornography based largely on research done at Carnegie Mellon University. Elmer-DeWitt's pitch to his editors was compelling enough to earn him the cover of the nation's most widely read newsweekly for July 3, 1995.

The subject was ripe for coverage. Members of Congress, prodded by the religious right and antiporn groups, were poised to legislate against what they perceived as the ability of minors to find reams of smut on the Internet. Civil liberties groups were trying to deflate Congress's concerns, saying they were overblown and that, in any event, parents, not government, should be the ones deciding how kids use the Net. What the debate lacked was some hard facts. And that, Elmer-DeWitt believed, was what the *Time* story would provide: coverage of a comprehensive university study detailing the availability of porn online.

The problem was that Elmer-DeWitt had been duped. The "Carnegie Mellon study" that *Time* brought to international attention with its exclusive cover story, calling it "exhaustive" and "significant,"[1] was actually a severely flawed research project by a dissembling undergraduate named Marty Rimm. Notwithstanding the fact there was a good deal of pornography online, Rimm's research methods, it turned out, were a sham.

Most remarkably, *Time* repeated Rimm's claim that 83.5 percent of the images posted on Usenet newsgroups on the Internet were pornographic. Rimm, however, was actually evaluating adult-oriented computer bulletin-board systems (BBSs) that were not connected to the Internet and that generally required a credit card for access, thus keeping children out. Claiming that there was a lot of smut on those BBSs was like saying that there is a lot of nudity in a hermetically sealed *Playboy* magazine.

Even before the *Time* issue hit the stands—with its sensationalistic cover of a wide-eyed, porn-engrossed child—free-speech activists and other Internet users were talking up a storm online about Rimm's research, the political implications of the story, and particularly why Elmer-DeWitt, a respected technology journalist, would give credibility to a study that seemed to have so many flaws.[7]

On the Well, an influential San Francisco–based BBS of which Elmer-DeWitt was a member, the exchanges were getting increasingly heated. And things only got worse on Monday, June 26, when the *Time* issue became available and was immediately brandished on the Senate floor and cited on national radio and television as proof that the Internet was awash in filth. Elmer-DeWitt became—like Senator James Exon, the Nebraska Democrat who first led the effort to criminalize indecency online, and Rimm himself—an Internet pariah, subjected to a seemingly endless barrage of stinging criticism.

"Don't you have a sense of self-respect? A sense of shame?" wrote Mike Godwin, counsel for the Electronic Frontier Foundation (EFF), who described himself as a friend of Elmer-DeWitt. "You've totally lost it, Philip. . . . Don't even bother talking to me any more."

Said Elmer-DeWitt: "This study was going to get covered whether I did it or not. Other newsweeklies were eager to run with it. It wasn't an easy story to write for a lot of reasons. I did the best I could."

"But Phil, you take a hefty swing at a loaded topic using a goddamn Wiffle-ball bat," said another Well-ite. "I can't believe you bought into those stats, or that you wrote them up like you did."

The response: "Let's take a breath here, OK? Yes, I wrote the piece. There was no gun to my head. It was not heavily edited. I have to take the heat."[3]

Participants in the Well discussion likened the piece to journalistic malpractice and demanded to know whether Elmer-DeWitt and *Time* still stood behind it. And the more Elmer-DeWitt tried to defend himself, the more dogged his critics became.

"There is not a minute's rest for Elmer-DeWitt," journalist Brock Meeks wrote, in a lengthy article about the exchange. "He is constantly hounded whenever he goes online."[4]

Some of this riposte, to be sure, was overblown. But there were also many thoughtful objections from a variety of Internet experts and denizens. Within a few days, Donna Hoffman and Tom Novak, Vanderbilt University professors known for their work on Internet commerce, posted online a point-by-point rebuke of the Rimm study and of *Time*'s handling of it. (Their critique was all the more damaging because, before Elmer-DeWitt published his piece, Hoffman had voiced concerns to him about it and he had discounted them.) Law professor David Post also quickly distributed online a detailed criticism of the Rimm study. And Godwin of the EFF exhaustively made the case against the *Time* story and the underlying study.

In the face of this reproach, Elmer-DeWitt felt compelled to concede publicly online that he had made mistakes. He acknowledged that his article should have mentioned criticisms of the Rimm study raised by seasoned experts like Hoffman. Indeed, if he had been under less pressure and had had "more presence of mind," Elmer-DeWitt wrote, he and *Time* would have asked an outside expert to review the study. Ultimately, he admitted to his critics that he had "screwed up" by failing to do the basic fact-checking that is the bedrock of good reporting.[5]

A few weeks later, *Time* published a follow-up piece by Elmer-DeWitt that was less candid than it might have been. It glossed over the way in which the magazine, in order to get an exclusive, had accepted Rimm's demand that the study not be shown to any outside experts or critics. And, according to Elmer-DeWitt, the concluding line that he wrote for the piece—"*Time* regrets its error"—was edited out.[6]

Still, the article could only be read as an admission that *Time* had been party to a hoax. Reassessing the prevalence of smut online, the magazine backed away from the 83.5 percent figure it had previously published and instead credited Hoffman and Novak's claim that porn represented less than one-half of 1 percent of all messages posted on the Internet. It admitted that "serious questions have been raised regarding the [Carnegie Mellon] study's methodology, the ethics by which its data were gathered and even its true authorship."[7]

What drove *Time* and Elmer-DeWitt to make these concessions? After all, it is not every day that a leading news organization is forced to discredit a cover story, especially one that had received so much attention. Nor is it common for a prominent journalist to admit publicly that he "screwed up."

Would the truth have triumphed even if the critics had made their case in more traditional ways—for example, by writing letters to the editor or publishing scholarly critiques of the coverage? Perhaps. But the Internet was integral to the counterattack. *Time*'s story would not have been discredited as quickly had the critics—specialists and lay people alike—not been able to coalesce online. Indeed, considering the momentum that was built instantly online, the resources that were pooled, and the comparative difficulty of creating such a critical juggernaut offline, it is doubtful that the story would, absent the Net, have been refuted as quickly, as publicly, and as unequivocally as it was.

Such a sustained effort could not have happened without some of the Net's unique code features—particularly its many-to-many interactivity, but also its openness, flexibility, and broad capacity, which led to the creation of an instant, burgeoning archive of evidence. With these features, a group of disparate individuals could share their concerns, swap vital information, find real expertise, come to conclusions, show their strength in numbers, challenge Marty Rimm's analysis, and ultimately inform the larger public and the media about the falsity of the cyberporn scare.[8]

The incident was a masterful example of digital activism and open debate in pursuit of truth. Individuals seized control of the flow of information. They did so in order to discredit a dangerously false report and thus influence public understanding of a vital social question—how to balance free-speech rights with the ability of the government to help parents protect their kids from certain adult materials.[9]

Taking on Big Media

Whether or not *Time* recognized the potential flaws in Rimm's study, there was no way it could have anticipated the spontaneous yet concerted response it would generate online. "This is the Internet's version of the O. J. Simpson trial," said Godwin.[10] The *Time* story was a milestone for the Net because it fueled cyberporn hysteria and encouraged Congress to pass the ill-fated Communications Decency Act (CDA). But as importantly, it was a turning point because it demonstrated how the Net would allow individuals to challenge the power of Big Media. Critics of the Rimm study may not have been able to prevent passage of the CDA; but they undoubtedly influenced journalists and editorial writers, most of whom wrote that the bill was draconian and unconstitutionally overbroad. The Supreme Court, in a unanimous decision, ultimately agreed.

It is perhaps logical that it would be a story about the Internet that would give its frequent users one of their first real opportunities to show what the medium could do. Since the *Time* incident, many other Internet controversies have prompted users to organize and try to reeducate the press and the public the way they did with the Rimm incident. Soon it will not be uncommon to find activists using the interactivity of the Net to transform public perception of national issues unrelated to technology. A misleading newspaper article or news segment about the defense budget or police brutality or affirmative action, for example, might prompt the same kind of sustained scrutiny and response that *Time*'s article did.

Already, leading journalists are having their feet held to the fire. Amateur media critics are using the Net to talk back to those cultural gatekeepers who have traditionally assumed that their interaction with the public was a one-way street. Jon Katz, a veteran journalist who became one of the first prominent online columnists, notes how different it is to work in an interactive medium where readers bombard journalists with responses to what they write. "The only thing I can compare it to," he says, "is being tied to the back of a car and dragged through the street."[11]

This ability of individuals to keep the media on guard is tremendously important because journalists are often the arbiters of the facts as we know them. Even if we accept that reporters are fallible and that writing can never be an entirely objective enterprise, we still rely heavily on journalists to sketch the contours of reality for us. We may be aware of the obvious editorial slants of a certain publication or author, but even the most vociferous media skeptics look first to the major daily newspapers and the evening news to find out "what happened." The opportunity, then, to hold the media accountable with objections and clarifications—or praise, for that matter—is one of the great values of an interactive medium like the Net. As Howard Kurtz of the *Washington Post* puts it, "the on-line feedback loop . . . helps put news organizations and consumers on a more equal footing."[12]

Of course, journalists are not the only members of society who use the press to shape our worldviews. Politicians, business leaders, entertainers, athletes, and religious figures all succeed in influencing our lives through their access to the media. Through a news conference, television talk show, press release, public event, film, radio interview, book, or op-ed, these individuals can air their views much more easily than the average citizen can (*pace* Jerry Springer and the other confession shows). This point may—indeed, it should—be obvious, and yet it is worth mentioning precisely because it may not be as true tomorrow.

That's because the Net not only allows us to dissect and criticize what is published, it lets us become publishers ourselves. We can do spin on the news or we can create the news. Even more than holding journalists accountable, this is the real way that Net users will shake the foundations of the fourth estate and the culture business writ large. With the ease and appeal of interactive media, we will increasingly become producers of information rather than just consumers of it.

Digital Auteurs

The writer A. J. Liebling famously quipped that "freedom of the press is guaranteed only to those who own one."[13] In recent years, as global conglomerates have consolidated their ownership of media outlets, Liebling's wry observation has seemed more apt then ever. In 1983, a few dozen corporations owned at least 80 percent of the market for television and radio programming, film, books, and magazines. In 1996, less than ten firms controlled around the same share of the market, and most of these firms were engaged in ventures together.[14]

But, even before the advent of the Internet, another trend was occurring as well. Though ownership of media outlets became more concentrated, the number of those outlets proliferated—because of new technologies like cable and satellite television and because of increasing specialization in industries such as magazine and book publishing. This simultaneous movement toward fewer owners and more choices has fueled a heated debate about how well we are being served by the media. Are we getting the news we need to be informed, responsible citizens? Are we getting diverse, high-quality information?

Optimists generally believe we are—or, at least, they believe things are improving. They see a cornucopia of information options where once the pickings were slim. A handful of television networks and major publishers are losing their dominance, they note, because of emerging technologies and changes in the marketplace. Not only do we have more channels to watch and more titles to read, but technical innovations have given us more choice as consumers. The birth of the videocassette recorder meant an unlimited storehouse of viewing options and the ability to control when we watch programs. Video cameras allowed us to be documentarians, artists, or even, as the Rodney King case made clear, public witnesses to injustice. To the optimists, even before the arrival of the Net, consumers in the media marketplace never had it so good.

> "The Internet puts the masses back in mass media."
>
> —Howard Rheingold

Skeptics, on the other hand, see the increasing concentration of media power as a threat to democracy, free expression, and civilized life. They maintain that the flow of unrestricted, quality content is inhibited because, no matter how many outlets there are, they are increasingly owned by a handful of megaconglomerates who care only about the bottom line.[15] Fear of alienating advertisers, executives, and shareholders, and a desire to appeal to the lowest-common-denominator audience, they say, means less diversity and risk in programming. (It's the Springsteen gripe about fifty-seven channels and nothing on.)

Until now, there were good reasons to believe the skeptics. But now along comes the Net and the potential for everyone to be a publisher. And suddenly it does appear that the balance of power could shift toward a more democratic equilibrium, that the "vast wasteland" of mass media might finally bloom with diversity and character. The control revolution promises, at least in theory, to give each individual or group the ability to disseminate speech far and wide without having to get permission from a Rupert Murdoch.

Liebling's old saw might still be true, but now owning a "press" is a possibility within almost everyone's reach. As author Howard Rheingold puts it, "The Internet puts the masses back in mass media."[16]

Renegade reporters like Matt Drudge, who broke the Clinton-Lewinsky story online in his *Drudge Report,* are already showing how this power shift might unfold. There are, to be sure, reasons to be concerned about the integrity of information in such an environment (as I will discuss in chapter 12). But there is also something undeniably novel and encouraging about the way that this great expansion in information sources might liberate us from the merger mania that has defined the communications industry in recent decades.

But, say the skeptics, how can the Net be an antidote to media concentration when it is still an exclusive medium, disproportionately accessible to those who are educated and wealthy? This is certainly true on a global scale. In scores of nations around the world, a basic telecommunications infrastructure hardly exists. There may be, for example, hundreds of people for each telephone line. (One African official reports that there are more phone lines in Manhattan than in all of sub-Saharan Africa.[17])

But in the U.S. and the rest of the developed world, telephone and television penetration rates are fairly high.[18] And the number of Americans on the Net has roughly tripled in the last three years. Inequality of access to the Net will likely continue to fade as computer prices continue to fall, as schools and libraries become wired, and particularly as the Net and digital television become integrated in the years to come. Today, Net access via TV is available for a few hundred dollars. Of course, as digital literacy and economic well-being become intertwined, inequalities relative to technology will remain. In the developed world, though, the ability to afford a Net connection will probably not be one of them.

A decade from now, in fact, anyone with access to basic communications technology should be able to enjoy or contribute to the diversity of the Net, which already is a bracing alternative to the conformity of old media. Web 'zines, online newsletters, and email lists are ubiquitous. (In late 1998, there were an estimated 90,000 public email lists, 30,000 Usenet discussion groups, and 23,000 Internet Relay Chat [IRC] channels.[19]) Academics are putting specialized journals online and other professionals are using electronic databases to find the vital facts of their trade. On the Net, one can locate almost anything that is commonly found in print—from newspapers to yellow pages to the Koran—and much more.

The Digital Freedom Network, for example, specializes in publishing writings that are censored by states such as China, Cuba, and Algeria. As its web site explains, it "provides dissidents with a global audience while eluding government control."[20] Similarly, the *Bolt Reporter,* an online newspaper for teenagers, has a special section in which it prints stories banned by school newspapers.[21]

Beyond thwarting censorship, the Net gives individuals unprecedented opportunities to share all sorts of creative expression. Visual artists are building virtual galleries in which to show their work. Cartoonists are syndicating their strips online instead of in newspapers. Musicians are putting their compositions on the Internet for others to hear.[22] As bandwidth expands and technologies improve, digital auteurs might even go head-to-head with the Disneys of the world—creating a wide-open market for cheap video distribution. You won't need a radio station and broadcast license to be a deejay or a TV station to be a newscaster. All you'll need are the tools to get your message on the Net.[23]

From the standpoint of democracy and freedom of speech, this is the Net's richest potential feature: individuals will exercise more control over the flow of information, and over the way that society understands issues and, ultimately, itself. Put another way, if information is, as some claim, our most important commodity, what could be more egalitarian than placing the means of production in the hands of individuals? So long as we can keep powerful entities from exercising too much influence over the Net and also find reliable information—two crucial provisos that I will address shortly—the payoff will be tangible.

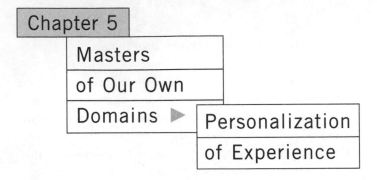

Chapter 5

Masters of Our Own Domains ▶ Personalization of Experience

There's a memorable scene in the classic 1960s film *The Graduate* when Dustin Hoffman's character, Benjamin, receives a single word of career advice from an older family friend. "Plastics," says the man, leaving Benjamin somewhat bewildered. If the same film were being made today, the mantra for young Benjamin to mull over might well be "personalization."

Recognizing individuals' desire for control, companies are tripping over themselves to give consumers the ability to personalize experience. Thus we see a bevy of products and marketing campaigns with the prefix "my"—My AOL, My Yahoo, My Netscape.[1] A prominent icon on Microsoft's desktop says My Computer and there are folders entitled My Documents, My Files, and My Briefcase. (Who else would they belong to?)

Personalization, though, is more than just a marketing fad. It's a concept that encompasses many different types of personal control. Increasingly we can use the interactivity and flexibility of the Net to customize our intake of information, our products, and even our social interactions. We can, in a sense, become masters of our own domains.

Personalization can play an important role in helping us to deal with information overload—from TV and radio, newspapers and magazines, phone calls, and now email and the web. David Shenk argues in *Data Smog* that this deluge of stimuli threatens to make us less well-informed and more stressed.[2] Yet already we are figuring out ways to manage the rising tide of data. Most effective will be filtering tools that allow us to screen out information we don't want.

One of the primary areas in which this screening will occur is in our intake of news. Rather than having editors and producers determine what we read, hear, and watch—as we do with newspapers or television—we can use the interactivity of the Net to gather just the material we find interesting. Preliminary studies of Internet use show that experienced users are increasingly doing so.[3]

To test this out myself, I recently subscribed to a couple of the leading news personalization services that are available online. These services let you choose topics that are of interest to you, and then they send headlines and summaries of stories about those topics to your email box each morning.

One service, called Newspage, allowed me to choose from more than 2,500 topics. There were general categories like Business Management, Healthcare, and Media and Communications. Then there were subcategories: within Media and Communications, for example, there were topics like Motion Picture Industry, Television & Radio, and Advertising & Public Relations. Finally, there were sub-subcategories: within Television & Radio, I chose U.S. Cable Regulatory Issues, Public Television, and Interactive Television, from more than two dozen options. Among the 27 other topics I selected were Computer Life, Intellectual Property, and something called Downsizing, Rightsizing & Smartsizing.

Another service I subscribed to, Infobeat, had fewer categories, but I was able to get an array of other personalized information: a weather update for my area, sports scores, a horoscope, listings for what guests would be on the late-night talk shows, even reports on ski conditions at mountains nearby.

These services—aptly described by MIT professor Nicholas Negroponte as *The Daily Me*[4]—gave me tremendous control over my news flow. Each day, I skimmed the headlines and summaries that were sent to me, and if I was inclined to read more I would click on a link to get the full text. Compared to radio or television, where I obviously had no say over the news I received (other than the ability to change stations), this was an empowering way to learn about the world. It seemed more efficient than reading a newspaper or magazine, since I didn't have to scour and search to find the news that was of interest to me.

If the idea of personalized news seems familiar, it may be because of *narrowcasting*, a media trend of recent years that has led to a surge in magazines and television shows aimed at increasingly specific audiences. Producers create content for a particular niche of the population—say, retirees who like to travel—partly to satisfy the desires of that audience, but even more to cash in on highly targeted market research and ad sales. There is, however, an important difference between personalization and narrowcasting: in the latter case, the individual still is receiving information packaged by someone else and it may arrive whether one requests it or not. What's novel about personalization, by contrast, is the ability of individuals to decide what information they receive and how they receive it. From my own experience, I can confirm that there was something quite rewarding (and ego-pleasing) about this new level of control over experience.

Since the earliest experiments with networked computers, the ability to get personalized news has been one of the goals that futurists have raved about most. Part of the value in this filtering is giving individuals information they might not otherwise get. Daily newspapers, according to one estimate, use only about 10 percent of the information they gather each day; readers, in turn, read only about 10 percent of the paper. Of all the potentially valuable material that a newspaper assembles, then, only 1 percent of it goes to the reader, and it's unlikely to be the 1 percent that the reader most wants.[5] Why not put as much information as possible online and let the reader choose?

> Personalization means newspapers will generate a million editions— each an edition for one.

Today, that concept is becoming a reality. In addition to services like Newspage and Infobeat, almost every major paper in the world—and some that aren't so major—has a web site where its contents can be navigated selectively by a reader with a simple click of the mouse.

Instead of starting with the front page and flipping through pages and sections, you can choose Arts or International or Editorials and go straight there. You can also read *Le Monde* if you live in Santiago, Chile, or the *Detroit Free Press* while you travel abroad in Tunisia.

Soon, major newspapers like the *New York Times* and the *Washington Post* may follow the lead of Newspage and Infobeat and ask you if you want to customize your daily paper. What would you like to see as your lead story? Are there areas about which you'd like to have extra content? (Already, the web sites of these dailies provide more in-depth coverage than the paper versions do.) Are there topics that you would rather not read about? They can be excluded from your personal edition. And how would you like to receive the news? In your email box or at your doorstep with the familiar smudgy print? (The *Jerusalem Post* has experimented with a reader-customized newspaper that can be delivered via printers in the home that churn out what apparently looks like a regular newspaper.[6]) Instead of publishing one edition of the newspaper for a million readers, newspapers would begin to generate a million editions—each an edition for one.

News is not the only thing that individuals are starting to personalize. "Mass customization" is the term that manufacturers use to describe the personalization of consumer goods. Consumers use computers and the Net to give vendors precise information about what they want to buy. Coming to a mall near you, for example, is a three-dimensional body scanner that will take your measurements for clothing more accurately than a Savile Row tailor ever could. The data might be transmitted to a company like Levi Strauss, which would custom manufacture a perfect-fitting pair of jeans and ship them directly to you. Even traditional outlets such as Brooks Brothers are experimenting with computer-assisted mass customization—anything to get that perfect fit.[7]

The potential goes beyond clothes. A company called MySki allows customers to design their own skis on the web. Select the type of skiing you like to do, enter your skill level and vital statistics, and then pick a color and logo. The site even shows you a graphic image of what your new skis will look like before they are made by hand and shipped to you.[8] Mass customization can save manufacturers money, since they should have little or no overstock. And since the customization is automated, the cost to the consumer should be lower than when getting something custom-made was a luxury done by hand.

Welcome to Your Life

Beyond personalizing information and products, the Net also gives individuals the ability to redefine their work and social spaces. Employment, for example, is changing as more individuals can choose to work out of their homes. This may mean telecommuting full time, or being able to log into your office's computer system while you care for a child, or taking a vacation that's twice as long because you can work remotely for a few hours each day. It might also mean starting your own business or finding a new line of work. A recent study of the American economy describes a "free-agent nation" comprised of 25 million workers who are self-employed or independent contractors—thanks in large part to new technologies such as the Internet.[9]

Some people can even make a living selling goods and services directly online. As President Clinton noted in 1997, "It will literally be possible to start a company tomorrow, and next week do business in Japan and Germany and Chile, all without leaving your home, something that used to take years and years and years to do."[10]

As rewarding as these work opportunities can be, the possibilities are even more far-ranging when it comes to using technology to change one's social life. Traditionally, friendships and acquaintances have been structured by physical proximity. Outside of family relationships, we generally meet people because they are our neighbors, classmates, coworkers, or colleagues in some local organization (a church, social club, or advocacy organization). The Net, however, gives individuals the opportunity to extend their social network in a novel way. You can spend more time communicating and sharing experiences with others regardless of where they live. You can form online relationships that are based on common interest rather than on the happenstance of geography.

These online communities, also known as virtual communities, are perfect gathering places for hobbyists and others with quirky or specialized interests. But they're also havens for anyone who relishes the opportunity to interact with others who are similar. For example, groups as diverse as fans of swing music, chemistry professors, and asthma sufferers now congregate online.

To those who haven't experienced it, the appeal of text-based online social interaction, particularly real-time "chat," may be hard to understand. Why not just go to a local bar or café, or start a reading group? Yet devotees maintain that online communities satisfy a genuine human need for affiliation with like-minded others.[11] Indeed, these associations suggest the possibility of whole new forms of social life and participation. Because individuals are judged online mostly by what they say, virtual communities would appear to soften social barriers caused by age, race, gender, and other fixed characteristics. They can also be valuable for people who might be reticent about face-to-face social interaction, like gay and lesbian teenagers, political dissidents, and the disabled.

People with autism, for example, often have trouble communicating in person, yet many say that they find the mediated nature of online contact to be a soothing alternative. "Long live the Internet," one autistic wrote in an online discussion, where "people can see the real me, not just how I interact superficially with other people."[12]

Even those who have little reason to fear face-to-face encounters may find themselves newly empowered by interactive technology. Something as simple as the widespread use of email lets individuals control social interaction in ways they could not before. It provides an alternative both to the formality of letter writing and to the occasional burden of real-time interactions, whether face-to-face or by telephone. With email, it is perfectly easy and socially acceptable to send a one-line message to say hello or ask a question. Basic as this may seem, it is nothing less than a new way for friends and acquaintances to stay in touch.

The Activist's Advantage

Social and political activists, like those involved with Radio B92, may have the most to gain from the Net's global reach, interactive speed, and personalization of experience. Organizers can create not just alternative news sources, but sites for community building around political issues. Consider the case of Htun Aung Gyaw, a Burmese dissident fighting the military government that rules his South East Asian homeland, known now as Myanmar. Htun does his plotting not in Rangoon but in Ithaca, New York. Though once a rebel fighter in the jungles and on the streets of Burma, now he is a graduate student at Cornell University. By day, he works as a book reshelver in the library. At night, he joins his wife and children at home and uses the Internet to convene electronically with other Burmese democracy activists around the world.

The virtual community of Burmese dissidents revolves around a few web sites, such as BurmaNet, that carry the most up-to-date information about the struggle, as well as background materials for newcomers and journalists. Participants gather regularly in chat rooms to express solidarity, reminisce about the homeland, and debate strategy. More confidential communiqués take place by email.

"We are weak," Htun says. "That's why we need high tech: they have an army; they have power; they have money. This is a new kind of warfare we are fighting, Internet warfare."[13]

Htun is not unique in using the new tools of Internet activism, which are the tools of grassroots community organizing transformed for the digital age. There are email lists, newsgroups, and web sites dedicated to almost every political conflict in the world, no matter how small.

Use of the Net by activists shows how the ability to control information and experience ultimately means the ability to create new forms of social life and political power. Sometimes solidarity may be the main benefit that is achieved, yet activists also credit the Net with helping them to achieve some concrete successes. A leader of BurmaNet, for example, says that online organizing among college students helped convince a major Western company to pull out of Burma.

An East Timor activist adds: "The Net has been so intrinsic to organizing in the US and internationally for the last few years that whatever successes East Timor's solidarity movement has had cannot be considered otherwise."[14]

Domestically, activists have used the Net to organize on almost every conceivable issue, particularly those that have to do with cyber-rights. Privacy, in particular, has been a hot-button issue. Many proposed corporate plans to make personal data available have been thwarted by online organizing.[15] Government censorship of speech has been another area of heavy activity. Net activist Shabbir Safdar says that during the campaign against the Communications Decency Act, between 65,000 and 100,000 people were reading the alerts sent out by his grassroots group, Voters Telecommunications Watch. In the spring of 1995, Safdar and Jonah Seiger of the Center for Democracy and Technology organized an online petition against the CDA. The petition, which was 1,500 pages long and listed 112,000 signatories, was printed and given to Senator Patrick Leahy, who in turn brought it to the Senate floor during the debate over the CDA.[16]

One form of activism that can benefit substantially from personalization is the effort to preserve the language and traditions of vanishing ethnic groups. Where the mass media provide homogeneity, the Net allows for a flourishing of diverse subcultures. In Hawaii, a movement is underway to save the Hawaiian language, which has become increasingly foreign to native inhabitants since missionaries began their effort to abolish the local argot more than a hundred years ago. A large part of this effort consists of computer education for students and online networking that strengthens ties between native speakers and those who want to learn.[17]

Perhaps the most novel form of personalization that the Net allows is the ability of an individual to experiment with identity. Online you can step outside of an assigned role—man/woman, child/adult, black/white, gay/straight—and try on another. It is not uncommon in chat rooms to find, for example, an older man posing as a teenage girl or a heterosexual woman trying out life as a lesbian. Some observers find this identity play disturbing, particularly because a kind of fraud may be perpetrated against the unwary. But experienced users know to be guarded about any online statements regarding identity, and psychologists generally believe that experimenting online is a safe and productive way for people to explore alternative viewpoints and experiences.[18]

Again, it is the text-based nature of most cyber-interactions that allows for this masking of the physical self. This is one reason, notwithstanding the Net's increasing capacity to carry video, that text will continue to be an important mode of virtual interaction. Additionally, role-players will adopt personas online—called "avatars"—that include visual and audio characteristics of their choice, thus letting individuals continue to manipulate how they present themselves to the world.

I will return in Part Three to personalization of news and social environments, looking at some unexpected effects it might have. Next, though, I consider another new form of individual control: the ability to bypass middlemen.

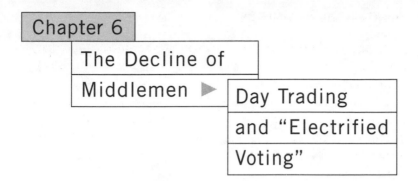

Chapter 6

The Decline of Middlemen ▶ Day Trading and "Electrified Voting"

As I write, I'm shuttling nervously between this word-processing document and a new favorite web site, and all I can think is: Uncle Max, I'm sorry.

You see, my uncle is a stockbroker—my broker—and the web site that I'm eyeing is an online brokerage account that I've opened without telling him.

What drew me to the site initially was the ability to get instant updates on the value of my small, yet growing, cache of securities. It's information I could never get before on my own. Beyond just quotes, I can get access to research from major investment houses, up-to-the-minute news about different companies, and even corporate filings. I also can look at graphs plotting the performance of any stock in scores of different ways.

And then there are the chat forums, where die-hard day traders with handles like "webinfopro," "fallenrise," and "cyberfortune2000" exchange tips on the hot stocks of the day. Or the hour. Or the minute.

But right now, there's something on the site even more alluring than all that information. It's a big, clickable Send Order button that is going to let me do something I've never done before, something I wouldn't even have cared to do: buy a stock without consulting Uncle Max—or paying him his commission.

In a flash, I've sidestepped the establishment and taken on the market by myself.

▶

Online investing is the control revolution made visible to the financial world. As Securities and Exchange Commissioner Arthur Levitt says, "Millions of new investors have taken advantage of the unprecedented access and individual control the Internet provides."[1]

They are using the interactivity, speed, and digital capacity of the Net to make decisions that were once made by others (brokers like my Uncle Max). They're exercising more control over resources that were once overwhelmingly controlled by powerful players (major institutional investors). And they're saving money, since commissions for online brokerages are much lower than traditional brokers, even discount brokers.

Many online investors, no doubt, are people like me who never paid attention to the market until they were given the power to manipulate it directly. By the end of 1999, an estimated 10 million investors will be trading online. Already there are more than one hundred online brokerage services and trading on the Net accounts for 25 percent of all retail securities investing. The returns for dedicated day traders—those who buy and sell furiously for hours at a time—are the stuff of fairy tales: the guy who went from making $2,000 a month as a flower salesman to earning $175,000 a month as a day trader; the scuba-diving instructor who made $27,000 in a single day.[2]

The effect on the market, experts agree, has been major. Many observers see the actions of day traders as a key factor, perhaps *the* key factor, that has driven Internet stocks sky high. It's how a company like Amazon.com, the online seller of books and other products, which isn't expected to make a profit until 2003, increased in value tenfold in 1998.[3] And it's how tiny companies like e-Bay and theglobe.com shoot up in price almost that much on the very first day they offer shares to the public.

Online investing shows how individuals are taking power from intermediaries.

Day trading is a perfect example of a theme that has actually been central, yet unstated, in our discussion thus far. In looking at the ways in which the Net allows individuals to assert themselves and take more control of their lives—by becoming critics and publishers, by personalizing news and products, and by establishing new communities and identities—we have been charting the decline of certain middlemen in society.[4]

Disintermediation is the somewhat ungainly word that is used to describe this circumventing of middlemen. Generally, it is associated with the ability to engage in commerce directly without brokers, retailers, and distributors. But the concept can be usefully expanded to describe the way that technology allows individuals to bypass editors, educators, and other gatekeepers who stand between us and whatever it is we seek. The control revolution allows us to take power from these intermediaries and put it in our own hands.

In the *Time* cyberporn case, for example, people with firsthand knowledge of the facts described by Philip Elmer-DeWitt disseminated information directly to the online public, rather than relying on an established publisher to do so for them. Newspage and Infobeat allowed me to jettison editors and make my own decisions about what was newsworthy. And the activists at Radio B92 evaded the ultimate gatekeeper—a state censor backed by military force.

Still, the most common examples of disintermediation can be found in the commercial realm. In this sense, online investors are archetypal: Why rely on a middleman, they ask, when I can do it all myself? Except they can't do it *all* themselves. Even though they may not need a traditional broker to place an order for them, they generally rely on online brokerage services to execute trades for them.[5] And even the stock markets—the New York Stock Exchange, Nasdaq, and so on—continue to function as intermediaries of sorts, since they bring together a buyer and a seller of a security.

More accurately, then, day traders are not getting rid of middlemen all together. They are merely peeling back a layer of intermediaries who are no longer necessary—who don't add value to the transaction any more. This is usually what people mean when they use the term disintermediation.

Amazon.com for example, is frequently cited as an example of disintermediation. And yet, although Amazon allows individuals to bypass many middlemen in the book business—most notably, local bookstores—it is itself an intermediary, albeit a new kind of digital middleman. The same is true of a company like Virtual Vineyards, which uses the web to sell wine, mostly from small winemakers, directly to customers all over the world. These companies are not in the business of publishing books or producing wine, let alone writing words or growing grapes.

But, the same way that online investing gives individuals more choice and control, and saves them time and money, Amazon and Virtual Vineyards give individuals new opportunities. Not only can you buy a book or bottle of wine at your convenience, but these services have capitalized on other code features, most notably interactivity, to give consumers something more. Potential book-buyers can read reviews from professional critics and other Amazon customers, post their own reviews, and read interviews with authors. If they like books by Philip Roth, they can find out what other Roth fans have purchased. On the Virtual Vineyards site, oenophiles can review the available wines by price, region, type, or vineyard. They can get information about body, intensity, and acidity; ask questions of a sommelier; and learn how to match wine and food.[6]

And then there are sites that bolster consumers by giving them new kinds of information. PriceScan, for example, uses agent technology to compare prices instantly for a product at a number of different online stores.[7] The prices are live quotes and you can follow through with a purchase. I logged on and asked PriceScan to find me the cheapest available price for Natalie Merchant's album *Ophelia*. It searched half a dozen sites on the web that sell CDs and returned a few seconds later with prices ranging from $11.89 to $14.08. (Not surprisingly, some vendors won't let the PriceScan agent get access to their sites—threatened, no doubt, by giving consumers too much control in the market.[8])

There are, of course, purer examples of disintermediation. Bylines, a web site maintained by Pulitzer prize–winning journalist Jon Franklin, for example, offers customers the writings of professional authors for small payments that they make using a credit card. "The idea is to cut out publishers, printers, advertisers, and glitz," says Franklin.[9] The works cost anywhere from a few cents to $2.50 for a book and can be read online or downloaded for printing. (Of course, an even purer version might have the authors selling their works on their own web sites, rather than on Bylines.) Other examples of near-complete disintermediation include major computer companies like Dell, which sells its own products directly to customers over the web. And then there are online auction sites like e-Bay, which allow individuals to engage in commercial transactions on an almost one-to-one basis (e-Bay itself is the last remaining intermediary).

Whether we're talking about a total absence of intermediaries or the demise of certain middlemen and the rise of others, the point is that the Net can give individuals more leverage as consumers and ultimately more control over commerce. There are, of course, people who think this may be a mistake. Many market observers, not surprisingly, believe that day traders are destroying the integrity of the stock market by ignoring fundamentals and driving up stock prices without any sense of what they're really worth. Whether these claims have merit is a question I'll return to in Part Three.

We should recall, though, that disintermediation is occurring in areas other than just commerce. Students engaged in learning online may try to do without traditional educational institutions and teachers. Grassroots journalists will sidestep major media. And, perhaps most profoundly, citizens may try to circumvent their elected leaders.

Disintermediating Politics

Politically, disintermediation is a close cousin to decentralization—the process of moving decisionmaking power from central to local authorities. In representative democracies, governments are run by elected officials and career employees who are intermediaries between the people and their collective resources and interests. We rely on these political middlemen to collect revenue, through taxes and other receipts, and to figure out how it should be spent to provide us with the benefits of life in a modern society.

In recent decades, calls for decentralization have come from both the left (for example, local control of schools) and right (almost the entire Republican agenda recently, with the exception of crime and defense). De-

centralization has been evident in the "devolution revolution," where social welfare tasks once handled by the federal government have been transferred to the states.

In short, decentralization attempts to bring decisionmaking power
closer to the individual: whether it's from the federal government to
state, state to county, county to munici-
pality, or municipality to community
board. In a sense, it embodies the collo-
quial lament that we should "toss the
bums out" and run government ourselves.
The logical conclusion from this is that
politicians are simply middlemen who deserve to be disintermediated.

Once, such an idea would have seemed not just audacious but impossible. Yet since the advent of interactive, networked computers, the
possibility of direct democracy—of citizens controlling the political
system directly, rather than through elected representatives—has been
very much in play. Ross Perot popularized the idea of electronic democracy in his 1992 campaign, when he spoke of wanting America to gather
in interactive town hall meetings to tell him what to do once he became president. The history of direct democracy, though, is as old as
democracy itself. The idea of the people (the *demos*) ruling themselves
was born in fifth century B.C. Athens, where political decisions were
made not by elected representatives but by citizens gathered in assembly (privileged adult male citizens, that is). These Greek city-state
inhabitants saw it as their duty to govern themselves. It was a responsibility that included not only coming together almost weekly to debate and vote on proposals but serving for as long as a year in
administrative posts.[10]

Even direct electronic democracy has a pedigree that goes back more
than half a century. In 1940, the futurist Buckminster Fuller wrote that
democracy was in a shambles and needed to be modernized "to give it
a one-individual-to-another speed and spontaneity of reaction com-
mensurate with the speed of broadcast news." To achieve what he called
"electrified voting," Fuller had a plan: "Devise a mechanical means for
nation-wide voting daily and secretly by each adult citizen of Uncle
Sam's family."[11]

> Once, bypassing politicians would have seemed not just audacious but impossible.

As far back as the early 1970s, researchers have conducted experiments in teledemocracy, giving citizens the opportunity to vote electronically from their homes.[12] Today there are a number of organizations and pundits who support Fuller's vision of direct democracy.[13] More recently, some cyber-pundits have upped the ante, questioning whether the rise of the wired nation doesn't make representative democracy, or at least the federal government, obsolete.[14] Citizens, they point out, can use the Net to express their preferences instantly and directly—not just in elections for candidates, but on specific policy questions like how we should balance the budget or where a hazardous waste dump should be located. Representatives might be kept on, but rather than making decisions independently they would simply become tabulators of the people's will as determined by electronic voting in ongoing plebiscites.

Americans have had some experience with direct democracy via the ballot referendum process, which is common in California and other western states. Referenda are also frequently held in other nations such as Switzerland, where citizens can petition to override legislative decisions or to pass laws that wouldn't otherwise be considered.[15] But unlike these initiatives, which require a tremendous amount of organizing and effort, direct electronic democracy might be simple, instantaneous, and ongoing.

There is much to question about such a political system, as I will discuss later. For now, it is simply worth pointing out that the Net seems to make it a possibility rather than just a science-fiction fantasy.

Having explained the potential of the control revolution, I turn now to the turbulence it is creating, starting with opposition from institutional powers.

Part Two

Resistance

"Wherever there is power, there is resistance."

Michel Foucault[1]

In the annals of revolution, perhaps no motif is more familiar than the idea that power and resistance go hand in hand. The vanguard resists the established elite. The elite resists the ascendant vanguard. And this back and forth continues until one side or another expires.

This is no less true in the case of the control revolution: There is resistance to the individual power that technology makes possible, yet it is not always obvious. Some repressive states, to be sure, are responding with brute force. But most institutional powers, whether they are governments or corporations, are not reacting so bluntly. Rather, they are altering the code of the Net in subtle ways that may diminish the new personal autonomy. One reason their efforts may be effective is that code changes are likely to be perceived as technical and inconsequential, when in fact the opposite is true. Another reason is that institutions may succeed in creating the illusion of personal control.

Does this resistance suggest a ruthless desire on the part of institutional leaders to maintain power? Sometimes. But more often it reflects a clash of values, a break between old and new ways of understanding the capacity of the individual. Though most politicians and companies extol the new individual control, they often act instinctively in ways that deprive us of it. Whether such restrictions are ever justifiable is a question I will touch upon here and then return to later in the book. The main purpose of the following chapters is to show some of the ways that resistance to individual control manifests itself.

Chapter 7

An Anxious
State ▶ Controlling Speech, Secrets, and Creativity

If governments were Hollywood caricatures, the average nation-state today might be a cross between Arnold Schwarzenegger and Woody Allen: a neurotic powerhouse worried about its own irrelevance and eager to show otherwise.

It's not an easy time to be a state. Approval ratings are down. Budgets are tight. Competition is global. Add the fact that new technology may be changing the relationship of citizens to the state, and it's not hard to see why many governments are in an anxious mood. Computers and the Internet, to be sure, present them with opportunities to enhance their power through activities such as automation and surveillance. Yet these technologies also are forcing them to deal with many novel challenges.

Not all of these have to do with individual control. The ability of rival nations or terrorist groups to use the Net to launch a crippling attack on a government's information infrastructure, for example, is a major new threat. Still, in democratic legislatures and authoritarian strongholds alike, there seems to be a unique—if sometimes unspoken—concern about the ways in which individuals can use new technology to challenge established political and legal institutions.

Part of this concern stems from the seemingly unaccountable nature of online actions. With its global and potentially anonymous nature, the Net can frustrate government attempts to hold individuals responsible for their behavior. How can a prosecutor track down child pornographers who use "anonymous remailers" to distribute their materials while hiding their real identities? How can a court award damages to a defamed person when the libel was perpetrated online by an offender in a faraway land? How can the state protect consumers when it cannot even determine who owns a web site that is distributing fraudulent material?[2]

> Politicians and companies extol the new individual control, yet act instinctively in ways that deprive us of it.

Governments, of course, have always exercised authority over their citizens. And much of this sovereign power is a legitimate part of the social contract, the implicit agreement we make to live in a civilized world rather than a barbarous state of nature. We expect government to preserve order and liberty, punish criminal behavior, safeguard the rights of minorities, enforce contract and property rights, and protect national security.

The control revolution, however, complicates things. Most obviously, governments will sometimes perceive the new individual control as a direct threat to state power and will forcefully resist it.[3] Less apparent and potentially more important, however, is the fact that individuals are acquiring abilities that neither they nor the state fully understand yet. Thus, even when government legitimately seeks to act in the public interest, it may wind up unjustly limiting the individual autonomy that technology allows.

Controlling Speech

The most unequivocal examples of resistance are found in governmental responses to the new ability of individuals to control the flow of information. This capacity breathes life into the lofty words of Article 19 of the Universal Declaration of Human Rights, which states: "Everyone has the right to freedom of opinion and expression; this right includes freedom to hold opinions without interference and to seek, receive and impart information and ideas through any media and regardless of frontiers." With its interactivity, speed, and global reach, the Net would seem to make that ideal possible.[4]

Unfortunately, many nations of the world have only contempt for these words and for the larger principles of freedom of expression, association, and thought that they represent. Astonishing as it may seem at the dawn of the twenty-first century, many heads of state still tremble at the prospect of unconstrained citizen dialogue. They know that unfettered speech can shape and transform individuals' expectations, giving them a renewed sense of the possible. Once people in isolation learn what life is like elsewhere, they may look more critically at their own circumstances—and they may not like what they see. States will provide other rationales for curbing speech, but the most blatant censorship occurs because autocrats simply do not want their own subjects to know that the abject life they lead could be otherwise.

The sharpest concern about the free flow of information can generally be found in Asian countries, where there often is a strong tradition of mistrust of individualism. The Net, with its potential to empower individual speakers, therefore presents a special threat. Authoritarian nations in this region have attempted to limit their citizens' access to materials on the Internet, fearing that such exposure will create fissures of dissent.[5]

The Chinese government, for example, has assiduously tailored the information its citizens can find online. Internet users in China are required to register with the police and are prohibited by law from spreading information that "hinders public order."[6] To exercise greater control over the content that enters the country, the government routes Internet traffic through a few electronic gateways in major cities. A 1996 law requires all Internet users to connect through these gateways. ("No group or individual may establish or utilize any other means to gain Internet access," the law says.[7]) Among the web sites that reportedly have been routinely blocked are not just those dealing with sensitive matters like human rights and Tibetan independence, but the sites of the *New York Times, Wall Street Journal,* and CNN.[8]

China's effort to stem Internet use by democracy activists has begun to result in some widely condemned prosecutions. In January 1999, Lin Hai, a thirty-year-old Chinese software entrepreneur was sentenced to two years for "inciting the subversion of state power." And that was simply because he supplied email addresses to dissidents abroad who published a pro-democracy journal on the Internet. The government also confiscated Lin's computer, modem, and telephone.[9]

> "The Internet must be controlled."
>
> —a Vietnamese official

When it comes to innovative attempts to control the Internet, few nations can match Singapore. The government of this tiny yet economically prosperous island nation has both encouraged citizens to get online and forced Internet service providers to block certain web sites that it deems inappropriate. All providers of religious or political content in Singapore must register with the state broadcasting authority, whose guidelines prohibit information that "tends to bring the government into hatred or contempt, or which excites disaffection against the government."[10] Singapore reportedly employs a staff of at least eight censors who surf the Internet daily looking for objectionable material.[11] (Recently, I received an email out of the blue from a Singaporean government official, who told me that she had heard a radio interview I'd given to the BBC in which I mentioned Singapore's practice of restricting access to certain web sites. After confirming that Singapore does restrict access to at least one hundred sites, the official invited me to visit the government's web site "for a better understanding of our role.")

The efforts of China and Singapore to censor the Net get attention around the world because of their important roles in global affairs. Yet things appear even tighter in some smaller and less conspicuous Asian countries such as Myanmar. In response, no doubt, to the valiant efforts of digital activists like Htun Aung Gyaw (whom I mentioned in Chapter 5), Internet use is completely illegal in Myanmar and all interactive technology is tightly regulated. Recently, a supporter of the Burmese pro-democracy movement reportedly died in prison, where he was being held for the crime of using a fax machine without a license.[12] Other nations in Asia, including Malaysia, Indonesia, South Korea, and Vietnam, have also censored the information their citizens can find on the Internet.[13] A Vietnamese official explained his government's view most succinctly: "The Internet must be controlled."[14]

Elsewhere in the world, similar reactions to the Net can be found. Saudi Arabia agreed to give its citizens access only when it was able to install filters that would remove all "objectionable" material.[15] Bahrain, Jordan, and Kuwait have placed similar restrictions on individual use, the latter stating that Internet service providers would have to block not only pornography but "politically subversive commentary."[16] The Iranian government sponsors an online network with chat rooms that allow dialogue—but only between two subscribers at a time.[17] Apparently, allowing three people to speak openly is too threatening.

Would Iran perceive the same threat if three people spoke on the telephone or if they met in an open market? Probably not. Though many Asian and Middle Eastern nations routinely censor TV, radio, books, and public speeches, something about the Net seems to produce a more pronounced sense of anxiety among government officials. Maybe it is the volume of the information that can be transmitted. After all, confiscating a few subversive pamphlets or videotapes at the border is a cinch compared to trying to stop the flow of millions of bits of information a second, which is like trying to hold water in a strainer. Whatever the precise nature of the threat, the changing balance of power seems to make many governments uneasy—and not just authoritarian ones.

Western nations have also taken steps to curb online speech, though the rationales differ from those put forth in the East. Rather than guarding against direct criticism of the state, officials in Europe and North America speak of the need to uphold public safety and morality, often with reference to the importance of protecting children from harmful or inappropriate materials.

These are legitimate goals. Yet, in practice, even well-meaning governments often go astray for a few reasons: first, because of their unfamiliarity with the technology; second, because of an exaggerated fear of the absence of hierarchical control over information; and third, because of a lack of understanding of how important political principles should be applied in this new technological context. As a result, blunt regulations are put in place and logic-defying prosecutions are pursued.

Sometimes, the driving concern of the state seems to be that someone needs to rein in the Net—if not government, then private actors. Consider the decision of law enforcement officials in the German state of Bavaria to prosecute Felix Somm, the German head of Compuserve, a major international online service (now owned by America Online). Somm was prosecuted in 1997 because a Compuserve subscriber used the service to obtain illegal pornographic materials from somewhere on the Internet. But Compuserve was not the producer of those materials and neither Somm nor any other Compuserve representative had reason to know how that one subscriber was using the service. Nevertheless, because Compuserve failed to prevent its subscriber from accessing the proscribed content, Somm himself was indicted for trafficking in pornography, including sexual materials depicting children and animals. (The contraband materials also allegedly included neo-Nazi propaganda, which is illegal in Germany and other European nations.)

This prosecution can only be described as perplexing. As a German law professor noted, "This is a crime of omission, and to have a crime of omission you need to have a legal duty to act. Felix Somm has no duty to control the Internet."[18] Nor, more importantly, did Somm have the *ability* to control the Internet, even if he had been willing to turn Compuserve into a proxy for the state.[19] In true Kafkaesque fashion, the prosecutors were holding him accountable for something that he basically could not prevent—and that he had no reason to know anything about. Following Somm's indictment, Germany passed a law to immunize service providers from such prosecutions. Additionally, as Somm's case was being tried, the prosecutors reversed course and decided that it would be unfair to hold him responsible for material about which he had no knowledge.

But it was too late. In May 1998, a Bavarian court found Somm guilty and imposed a two-year suspended jail sentence and a fine of nearly $60,000. Upon sentencing him, the judge said: "Even on the Internet, there can be no law-free zones."[20] In other words, if the German state could not halt the flow of prohibited material on its own, then it would threaten companies and even private citizens with criminal penalties in order to get them to do so. It would coerce citizens into policing the Internet.

How did it expect citizens to do this? Not by tracking down child cyberporn offenders and making a citizen's arrest, but by altering the code of the Net. In Somm's case, this meant having Compuserve establish a surveillance mechanism that would monitor every action of each user—a development that would wholly change the nature of the online experience, but that would also, hopefully, ensure that Somm would not be arrested again.

One wouldn't even need to be an Internet professional to qualify as a target for such a coercive scheme. In 1996, German authorities brought charges against Angela Marquardt, a twenty-five-year-old university student, simply because she had put an electronic link on her personal web page to a leftist online magazine called *Radikal.* The magazine allegedly had published articles on how to make bombs and derail trains.[21]

German police had already tried and failed to get local Internet service providers to block access to the site, which was maintained on a web server in the Netherlands. Their next best strategy, it seems, was to go after individual Internet users who pointed people to the site. Again, the goal apparently was to counter what the Net makes possible by forcing individuals to dismantle relevant code features—in this case, a link on a web site.

Ultimately, a German trial judge dismissed the government's charge against Marquardt, but only because of narrow inconsistencies in the prosecution's case. What was never really shown, as a result, was the futility of the government's action. Even if Marquardt's link to *Radikal* had been removed, anyone with access to the Net could still have gotten access to the site through dozens of other links. Moreover, whatever "dangerous" material existed on the site was undoubtedly available in print. The Germans were therefore prosecuting a university student simply for pointing the way to material that could probably be found in a local library.

In doing so, the Germans were not distinguishing themselves as uniquely inept. A year earlier, U.S. lawmakers, upon learning that bomb-making information could be found on the Internet, set out to pass a law preventing its publication online. Yet the same information was available not only in public libraries, but was actually being published by the U.S. government.[22] This, however, would not be the last time that U.S. lawmakers hastily subscribed to the view that the Net was out of control and needed to be tamed. And like the Germans, they have repeatedly looked to private actors to take on the role of tamers.

Much ink has been spilled (and many pixels filled) over the Communications Decency Act, which was signed into law by President Clinton in February 1996 and struck down by the Supreme Court sixteen months later. But there is still something revealing to see if we look at the CDA through the scrim of the control revolution—particularly as an example of Washington's discomfort with a communications medium that resists centralized control.

As is well known, the CDA made it a felony, punishable by a fine and up to two years in prison, to knowingly transmit "indecent" messages to anyone under eighteen years of age or to display "patently offensive" messages so that they might be available to a person under eighteen. Less well known, though, was the law's safe-harbor provision—an affirmative defense that would prevent individuals from being prosecuted if they took certain steps. The safe harbor prevented prosecution of Internet users who restricted minors' access to indecent material by establishing a technological means of checking the age of those who might receive their communications—for example, with credit-card verification or an adult password. In essence, then, Congress was asking all Internet users—whether they ran a commercial web site or just used email or chat rooms for their own benefit—to change the way they used the Internet by changing the architecture of online interaction.

> The Communications Decency Act was born not out of malice, but apprehension and confusion.

The legal problems with this scheme, as the lower federal courts and Supreme Court recognized, were numerous. For one, the wording of the CDA was so vague and overbroad that it might have covered not just hard-core pornography, but ribald literature, lewd conversations, and even safe-sex information. Additionally, the statute's safe-harbor provision was unworkable, the courts said, either because the technology did not yet exist to screen out kids or because the available blocking mechanisms would be too costly for noncommercial providers such as nonprofit groups and individuals. In attempting to protect kids, then, the CDA would put a frigid chill on free speech, vastly restricting the materials available to adults. The law therefore violated the First Amendment, the courts concluded.[23]

This ruling was certainly correct. Yet, with all the commentary about the CDA and its failure, what hasn't been emphasized is the degree to which this statute was typical of a kind of resistance to the control revolution—a resistance born not so much out of malice but out of apprehension and confusion. To understand this it is helpful to consider the landscape of free speech prior to the rise of the Internet, particularly with regard to minors' access to adult materials.

The case of Sam Ginsberg is instructive. In 1965, Ginsberg and his wife were running Sam's Stationery and Luncheonette in Bellmore, Long Island. Richard Coray, a young man, walked in and asked to buy a few adult magazines that were for sale in the store. Without asking Coray for a driver's license or any other identification to establish his age, Ginsberg sold him the magazines. Soon thereafter, Ginsberg was prosecuted by Nassau County officials and convicted for giving materials deemed "harmful to minors" to a child under seventeen. Coray, it turned out, was sixteen years old. At trial, Ginsberg was found guilty by a judge who determined that Ginsberg knew the contents of the magazines, had reason to know Coray's age, and had a legal obligation to prevent Coray from receiving the material. The case went all the way to the Supreme Court, which upheld Ginsberg's conviction.[24]

Today it is well-settled law that commercial intermediaries such as booksellers and movie-theater owners have an obligation to act as gatekeepers, keeping certain materials away from minors. The same goes for other types of middlemen like radio and television broadcasters, though they have a stricter duty to prevent kids from being exposed even to common vulgarities. This was made clear in a famous Supreme Court case involving the comedian George Carlin. KPFA, a San Fran-

cisco radio station that broadcast Carlin's "Seven Dirty Words" routine, was sanctioned by the government for airing the program in the
afternoon, when children might hear it.[25]

On its face this might seem unfair. If society decides that minors
shouldn't have access to certain kinds of expression, why hold KPFA
or Sam Ginsberg responsible for carrying out this mandate? Why not
go after George Carlin or the porn publisher for making the material
available in the first place? Or why not arrest a kid like Richard Coray
for taking the affirmative step of buying it?

The answer, it turns out, is that placing the burden on the middleman has its benefits for society at large. On the one hand, it is efficient, because there are fewer middlemen to police than listeners or
readers. And on the other hand, it is more protective of speech than
subjecting the speaker or publisher to liability, because adults still have
unrestricted access to the expression and the speech of individual speakers is not chilled.

In looking at the Internet, though, lawmakers found no equivalent
of Sam Ginsberg, no middlemen to deputize in the cause of protecting children from inappropriate materials. Congress's solution to this
predicament was rather heavy-handed: Why not force every Internet
user to be a Sam Ginsberg? Thus the CDA applied to every individual who transmitted indecent material online. We were all supposed
to play gatekeeper.

The U.S. government, in other words, was taking note of the new
individual control over information, and it was trying to counteract
it.[26] Yet it was not doing so the way China or Singapore would, by having the state itself block certain sites. Rather, like the German government in the Somm case, it was deputizing individuals to do what
it could not—forcing them to alter the code of the Internet to keep
minors away from indecent content. Only instead of targeting just one
citizen who happened to run an Internet service provider (which might,
at least, have been efficient), it was threatening everyone at once.

To those who write headlines or argue cases in the court of public
opinion, all this may seem far less significant than the fact that the
CDA could have sent a person to jail just for posting an X-rated story
on her own web site. What was even more significant about the CDA,

though, was that it attempted to compel a change in the form of a communications technology.

Though the government's goal of protecting kids was legitimate, its execution was ham-fisted—in large part because of its poor understanding of the technology it was regulating.[27] The openness of the medium was to be replaced with a set of roadblocks, barriers that individuals would be forced to erect to restrict the movement of their own words and opinions. Such a change in the design of the Net would have affected its use and impact on society, in ways that the government did not likely intend. It was like telling pedestrians they couldn't speak to one another in public without first checking the IDs of everyone on the sidewalk to make sure they were all adults; people might just decide instead not to speak there.

This kind of technological resistance to the control revolution is unlikely to disappear. Rather, it will become more refined as governments become more adept at influencing code without running afoul of constitutional limitations or public opposition.[28] One reason they may succeed is that protocols of digital communication traditionally have been hashed out by obscure committees of engineers and computer programmers; thus, we are not yet accustomed to thinking of technology design as a highly politicized sphere of life. And yet it is the very obscurity of code regulation that would allow the government to gradually and imperceptibly alter technology to achieve its aims without public scrutiny.[29]

One area in which this pattern of increasingly nuanced regulation is evident is in the controversy over encryption technology.

Controlling Secrets

Encryption is a process by which digital information is encoded so that it remains confidential. A sender of information—via email, say, or a wireless phone call—can scramble the plain text of a message so that only an authorized recipient can decode it. And information can also be permanently stored in encrypted form for security reasons. Encryption tools, then, are the locks and keys of the digital age. They allow us to maintain privacy and feel secure about our computerized information and communications.

This means more than just preventing embarrassing disclosures. Human-rights workers rely on encryption technology to transmit sensitive case information that, if intercepted, could lead to brutal abuse of victims and their advocates. Health-care professionals rely on encryption to keep secure medical records for patients. Encryption also allows Internet users to ensure that information has not been tampered with and to authenticate their identity through digital signatures. And it further facilitates commerce by permitting credit-card purchases to be made securely online and by allowing vendors worldwide to exchange payments securely.

Even before the modern computer, encryption was used to protect the confidentiality of information. In time of war, for example, a cryptographer might encrypt a message using a key that would be known only by an ally. Using that key, the ally could decrypt the scrambled communication. A master cryptographer who intercepted an enemy's encrypted message would use logic and deduction to try to replicate the enemy's key and unscramble his communiqué. (In 1775, George Washington's troops were dealt such a blow when a private strategic message was seized and deciphered by a British spy.[30])

In recent years, technological developments have changed encryption. Keys have become computerized. The single-key approach has been replaced by a widely available technology known as public-key encryption that uses two uniquely corresponding keys, one public and one private. (Users make their public keys known and retain their private keys. Alice can therefore encrypt a message with Bob's public key and know that only Bob can unscramble it. And if Alice encrypts the message with her private key, Bob can use her public key to authenticate that the message came from her.[31]) The string of numbers in each key has also gotten longer, growing from 40 bits, or digits, to the current state-of-the-art 128 bits for PGP (Pretty Good Privacy) encryption.

All this means that encryption is easier to use—anyone can download the latest encryption software from the Net—and much harder to crack. Indeed, individuals now have access to "strong" encryption:

encryption that is unbreakable or that would take the most powerful supercomputers a long time to break.

Before the widespread availability of strong encryption, there was always the possibility that remote communications would be intercepted and read by the state (or by private snoops). Though government was only supposed to eavesdrop on those who were engaging in illegal conduct, rogue officials could abuse that power, tapping the lines of law-abiding citizens—or, before the advent of the phone, seizing written communications. Strong encryption changes this, because even if unauthorized interception of an encrypted message occurs, the message will be incomprehensible.

This changes the balance of power between individuals and the state. It allows us to keep secrets from government.[32] For the individual, this may seem like a great boon. As crypto-advocate Hal Finney says, "encryption offers for the first time a chance to put the little guy more on an even footing with the big powers of the world."[33] But from the vantage point of law enforcement, it's a whole new game—one that requires individuals to give up some control, according to government officials.

"The looming spectre of the widespread use of robust, virtually unbreakable encryption is one of the most difficult problems confronting law enforcement as the next century approaches," says Louis Freeh, Director of the Federal Bureau of Investigation.[34] "At stake are some of our most valuable and reliable investigative techniques, and the public safety of our citizens."

Freeh and his colleagues at the FBI and the National Security Agency are particularly concerned about use of encryption by terrorists and other criminals. They fear that during the course of a time-pressed investigation, law enforcement will intercept an important criminal communication only to find it protected by strong encryption and thus unreadable. Prior to the availability of strong encryption, of course, a criminal might have tried to evade the cops. But the state could respond with its privileged investigative tools—most likely, wiretapping. Now, these government officials say, the upper hand has been

effectively taken from the state. Strong encryption means law enforcement can no longer get timely access to the plain text of messages.
The only solution, these officials say, is to allow the state to retain its
advantage.

The U.S. government has pursued this goal via a number of strategies, which have involved increasingly sophisticated code regulation.
Beginning in the early 1990s, the government attempted to prohibit
individuals from getting any access to strong encryption at all. To accomplish this, it developed its own weak encryption standard, known
as the Clipper Chip, and then tried to use its purchasing power, its influence with technical standards-setting bodies, and just plain coercion to force this lame standard on industry and consumers.[35] As with
the CDA, this was a rather blunt attempt by government to regulate
code.[36]

The federal government also has tried to limit access to strong encryption by preventing its export. Until late in 1996, for example, the
government classified strong encryption as a munition, making it a
crime to send or carry it out of the country, and since then it has placed
other limits on export of encryption. These restrictions are necessary,
the government has said, because of the large number of potentially
nefarious users of encryption abroad. But there are likely other mo-
tives at work, as well. Export restrictions affect product development,
because companies generally prefer to develop technologies for one
global market. Therefore, restricting export is a more indirect (and
more palatable) way for government to continue its efforts to limit
general use of strong encryption.[37]

Finally, law enforcement has called for modification of strong encryption tools so that government will have a reliable way to decipher
secure communications. "Law enforcement cannot effectively combat
society's most dangerous offenders if there is no way to promptly decrypt their criminal communications and computer files," the FBI's
Freeh has told lawmakers.[38] Specifically, the FBI has urged Congress
to require encryption products to be built with a "key escrow" system.

Key escrow would oblige computer users to deposit a copy of their
private encryption keys with a company that would act as an escrow
agent. Law-enforcement officials engaged in investigation of criminal

activity could get quick access to the plain text of messages by presenting a warrant to an escrow agent requesting an individual's deposited key. What they are asking for, FBI officials say, is no more than the access they get when a court issues a warrant to wiretap a phone line.

To date, key escrow is the government's most elaborate and astute form of code regulation in the area of encryption. On the one hand, it would require a wholesale revamping of public-key encryption products. On the other hand, it might be a political winner because law-abiding citizens would appear to have the full benefit of strong encryption.

As with the state's effort to protect kids from sexual content, the goal here, protecting public safety, is indisputably important. Yet is this particular code regulation rational and warranted? Or is it driven more by anxiety about a new degree of individual control—and by a misunderstanding of what the outcome of such code regulation might be?

How, for example, would the government's currently favored solution, key escrow, really work? Its purpose is to allow law enforcement to intercept the communications of criminals. But why would any criminal use the government's mandated key escrow system? Already, strong encryption products without escrowed keys are widely available internationally.[39] And if forced to use hardware with key escrow built into it, criminals might just encrypt the underlying plain text using a form of encryption not subject to escrow requirements.

At the same time, key escrow could jeopardize the communications of law-abiding citizens, since unauthorized parties might get access to deposited keys. As the federal government's National Research Council observed in a 1996 report, key escrow "by design introduces a system

> The government's effort to regulate encryption could have the opposite of its intended effect.

weakness . . . and so if the procedures that protect improper use of that access somehow fail, information is left unprotected."[40] Untrustworthy foreign governments, for example, might claim a right to keys escrowed by third parties in the U.S. And high-tech criminals would likely see key escrow agents as perfect targets for attack, the same way bank robbers prey upon armored cars.

In this light, key escrow might be less like wiretapping and more like requiring individuals to store an extra copy of all their most sensitive physical keys—house keys, safe-deposit box keys, even keys to their doctor's office—in one central storage place. Encryption, after all, will be routinely used to safeguard more than just day-to-day conversations; as noted above, it will authenticate identity, protect commercial transactions, conceal medical records, and so on.

In short, the government's effort to regulate code could have the opposite of its intended effect, diminishing individual security while hardly affecting criminals at all. Whether this is a necessary risk—or a risk no different from that involved in conventional wiretapping, as the government claims—is a question I will return to in Chapter 16. What's important here is to see the increasingly intricate ways in which the state may, in the course of legitimate pursuits, limit individual control without justification—and without meaning to do so.

Copyright is another area in which code-based responses to the new personal control may produce significant unintended results.

Controlling Creativity

Copyright, like most areas of the law, rests on a delicate balance of interests. Creators of original works are entitled to a limited exclusive right to copy or otherwise exploit their creations, whether they are novels or textbooks, snapshots or epic Hollywood films. The goal of this exclusive right is to give creative people a financial incentive to produce art, further science, and expand the boundaries of human knowledge.[41] Our legal system also recognizes, however, the importance of making creative works available for the general benefit of society. Copyrights therefore last for a fixed period of time, after which they fall into the "public domain"—and the public can then freely use them.

Additionally, the public domain is effectively expanded by the fact that copyright law gives users a bit of breathing room even when using protected works. The principle of "fair use," for example, allows a journalist (or anyone else) to quote a few sentences from a copyrighted book in order to critique it. It lets an English teacher pass out photocopies of a marked-up poem to a class of students, and it gives a rap musician the right to parody a well-known rock and roll song.[42] Beyond fair use, no one thinks that copying a *New Yorker* cartoon to put on your refrigerator is copyright infringement. In legal terminology, that's a *de minimis* use—such small potatoes that no one should care and generally no one does.

Copyright, in other words, establishes a careful equilibrium between the rights of owners of creative works and the rights of users. Digital storage of information, however, changes things, making it easier for anyone to reproduce and instantly distribute protected material. The Net in particular—with its digital, interactive, distributed-network architecture—has drastically changed the dynamics, and the economics, of copying. Previously, the costs associated with mass copying and distribution of a pirated work were fairly high—high enough, for example, that book publishers didn't need to worry much about people photocopying books and selling them on the street.

The Net, by contrast, seems to be a gigantic copying machine. As a result, many owners of works, particularly large commercial content providers, face a new threat.[43] The problem is not just the professional thieves who sell black-market videos, music CDs, and computer programs. Rather it is the large, anonymous mass of casual Internet users who may have few qualms about unauthorized reproduction of copyrighted material. From the perspective of an individual, it's understandably hard to see the harm in making a digital copy of, say, a cartoon of Bart Simpson and placing it on a personal web site.

To counter this new degree of personal control of creative works, large copyright owners and their allies in government—call them "copyright maximalists"—have tried a variety of schemes.[44] An industry-oriented white paper issued by the Clinton administration in 1995, for example, called for the strengthening of copyright protection in ways that could have challenged basic use of personal computers and the Net (for example, potentially making every web page accessed by a user a "copy"). But criticism by copyright scholars and activists caused most of the white paper's proposals to be rejected when they were introduced in Congress.[45]

Administration officials made similar efforts in 1996 at an important international meeting of the World Intellectual Property Organization (WIPO), where a copyright treaty was being renegotiated, but again came up largely empty-handed. Among the defeated proposals was an attempt to create a new right to protect the contents of databases: facts such as telephone numbers or baseball scores that would not be protected under existing copyright law.[46] In 1998, copyright maximalists did succeed in lengthening the standard term of copyright protection by twenty years.[47]

More significant than this conventional attempt to expand copyright law, however, have been efforts to change the code of the Net to protect information. If the problem is that digital information can be copied and distributed widely by anyone, copyright maximalists have theorized, why not create technologies to neutralize this ability?

A first step in this direction came with the development of digital watermarks, which allow owners of protected works to track down pirated copies of their work on the Net.[48] All the owner has to do is use a search tool known as a web crawler to locate works with a certain watermark. If a person displaying the watermarked material did not pay for it, then the owner can threaten legal action. Because such pursuits still rely on law enforcement, though, they may not be worth the effort. After all, what's a big company going to do, sue every kid who posts a copyrighted Bart Simpson image on his web site?

Cost aside, a company might not even know for sure that it would win such a lawsuit. The user might, for example, raise a fair use defense. As a result, large information companies are increasingly trying to protect their assets by relying on contract law instead of intellectual property. Anyone who has used a software program or gone online in search of proprietary information has undoubtedly confronted the ubiquitous computerized form agreements known as "clickwrap contracts." These contracts limit the ways that information can be used, going beyond the protection provided by copyright law. Usually, they contain hundreds of lines of fine print followed by an Okay button that you are supposed to click to signal your assent.

The reality, of course, is that almost no one reads all the details of these agreements. And even if people did, these form agreements provide no opportunity for users to negotiate the terms of the deal, which is usually an essential component of a binding legal agreement. Nonetheless, courts will likely enforce clickwrap contracts.[49] And this may create a deterrent effect. Users might think twice about copying a Bart Simpson image if they have a hunch that clicking Okay may have in some way prohibited them from doing so.

An even more sophisticated and effective tool for protecting information is known as "trusted systems." This technology regulates how digital information is used: how many times it can be viewed, whether it can be duplicated, and so on. It seeks to tip the balance of control over creative works from the user back to the owner.[50] The key to trusted systems is that enforcement is built into the code itself. The owner of the Bart Simpson image, in other words, can simply program the image so that it cannot be copied when it is displayed or distributed.

The ramifications of trusted systems are particularly stark when compared to copyright protection in traditional media. A newspaper publisher, for example, can only prevent me from photocopying an article by suing me, which is costly and inefficient. On the web, though, trusted systems allow a content publisher to program a site so that I can't print its articles or even cut and paste them electronically, even if I have paid to view them. Similarly, if I buy an album on compact disk from a local record store, there is no way for the record company to know whether I make a copy of it for myself or for a friend. But if I buy and download an album from an online music vendor, trusted systems are available that will let the music publisher monitor how many times I play that album, charge me for each listening, and prevent me from copying it.

Trusted systems, clickwrap contracts, and digital watermarks are all attempts to fight fire with fire—using one technology to negate what another technology makes possible. They aim to achieve what the law on its own has not been able to do: prevent unauthorized copying. Yet these technological solutions also require the law's endorsement to ensure that individuals use them. Information owners therefore persuaded the U.S. Congress in 1998 to enact a law prohibiting users of the Internet and other digital technologies from circumventing tools such as trusted systems—even from using technologies that *could* accomplish such circumvention.[51] This itself was a regulation of code, as it required individuals to use certain technologies rather than others.

Thus, just as the U.S. government has sought to alter the code of the Net to respond to indecency online and the power of strong encryption, it has also helped to structure the Net's code to prevent individuals from taking too much control of creative works. As in those other cases, the government has a legitimate interest in protecting copyright and achieving the elusive balance that promotes both creativity and public access to knowledge. (It should especially look after the interests of those individual artists and writers who still make a living off their own copyrights, or at least try to do so.)

What we must watch for again, however, is the way in which well-intentioned code solutions may restrict individual control more than is warranted, shrinking the public domain and even inhibiting important freedoms.

Trusted systems, for example, generally make no allowance for fair use, that aspect of copyright that is essential to freedom of expression.

> The increasing reliance on code regulation is necessary, yet perilous.

We might therefore be deprived of information uses that have always been considered legitimate, such as passing on a purchased copy of a work to a friend or copying an article for yourself. (A determined person, of course, might manually copy a few lines of electronic text. But digital copies could not be made, a factor that is crucial as we live increasingly in a world of sounds and images rather than text.)

Trusted systems also monitor information use in a way that could jeopardize our right to privacy with regard to what we read, listen to, and watch. Each time you look at a picture online that is protected by trusted systems, for example, a computer somewhere takes note of that action. Under the guise of making sure that protected works are not used in an unauthorized manner, trusted systems could become perfect surveillance tools.[52]

▶

In Part Four, I will discuss what to do about these quandaries. Here the purpose has primarily been to see how the conflicts over indecency, encryption, and copyright demonstrate the complexity of resistance to the control revolution. Some state actors may be driven by a real desire to block personal autonomy. But in the U.S. and other democracies, officials seem to be motivated more by a general anxiety about how they will achieve traditional governmental ends in a changing world. The increasing reliance on code regulation is necessary, yet perilous. It's not easy for anyone—lawmakers, technologists, civil libertarians—to know how such machinations will play out. And it is too easy for government to quietly manipulate code without our paying attention, potentially reducing the new individual control to a chimera of the real thing before we even have a chance to enjoy it.

Governments, moreover, are not the only institutions that may resist individual control—as the next chapter illustrates.

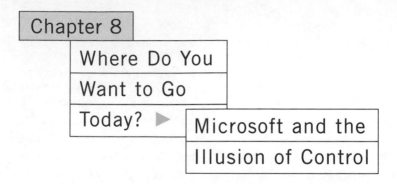

In early 1995, Microsoft began an international advertising campaign based on the slogan "Where do you want to go today?" The start of the campaign coincided with a nimble reorganization of the corporation by its chairman, Bill Gates.[1] Caught off guard by the Internet's huge growth, Gates shifted Microsoft's focus to this burgeoning new medium and to the promise of a networked world.

The new pitch fit perfectly. *Where do you want to go today?* Like the Internet itself, the motto places the individual at the center of the action. It asks you where you will go, what you will do, with whom you will interact. It triggers the imagination, hinting at the possibility of far-flung adventure—away from all the droning trivialities of life in the real world. The inclusion of "today" cleverly suggests a routine, like reading the morning paper, while also promising novelty. And the fact that the slogan is a question implies entitlement and service, as if the interrogator were some digital chauffeur waiting to whisk you off to your destination. Where do you want to go today, *sir?*

Microsoft's tag line is a savvy piece of marketing. But it is more than that. It is perhaps the perfect catchphrase for the control revolution, embodying the new, technologically enabled exaltation of individual will and unfettered choice. It is an ode to personalized control, borderless movement, seemingly unlimited opportunity—the spoils of the revolution.[2]

Microsoft is an important player in these developments, but it is not the only company with a slogan that celebrates personal empowerment. Its rival Netscape has run TV ads that end with the pitch "Netscape.com. The world according to you," while another competitor, Lotus, has borrowed a popular song by the band R.E.M. for use in its ads, which feature the lyrics "I am superman and I can do anything."[3] And for years, an AT&T ad campaign has similarly promised a world of new options, power, and mobility for the individual. In one segment, a man is pictured on a beach holding a computerized writing tablet. "Have you ever sent a fax from the beach?" the ad asks. "You will." Another shows a close-up of a wristwatch. "Have you ever installed a phone on your wrist? You will."[4]

In a sense there is nothing surprising about the breathless optimism of these slogans. After all, they are sales pitches. Yet it is worth asking: Does the corporate world really understand the personal empowerment that it is celebrating? Does it see the ways in which the control revolution will make the behavior of competitors and consumers less predictable? Does it understand that individuals can wield their new power in ways that could, quite simply, cut into revenue and business profits? (The complications could be especially acute for companies that rely on the *passive* behavior of the consumer.)

As Net users customize their information diet, bypass gatekeepers, and explore an array of new options, many companies—including online media

> Corporations may try to convince us that we are in charge even as they prevent us from taking control.

and commerce outlets—may have a hard time holding on to their most precious asset: the attention of consumers. It's an asset that translates into advertising revenues, direct sales, and licensing fees—the things businesses need to stay alive.[5] Ad sales, in particular, depend on the ability of content providers to capture and hold (and sell) viewers' attention. Yet the Net's interactivity works against top-down attempts to control the information to which people are exposed. If the TV remote control posed a challenge to television programmers and their advertisers, imagine the problems presented by the web, where users can, in theory, start where they like and go anywhere.

Considering this, it is worth asking: Why are Microsoft and other companies championing individual control?

The easy answer is the cynical one: These corporations are trying to deceive us. Following Rousseau's dictum that "there is no more perfect form of subjection than the one that preserves the appearance of freedom," they are trying to convince us that we are in control even when we are not—indeed, even when they are depriving us of it.[6]

There is, however, a more complicated and persuasive explanation. Sometimes businesses see individual power as a direct threat and actively seek to limit it. Yet more often, resistance occurs without grand designs. Like governments, corporations pursue what society deems to be legitimate goals: increasing market share, maximizing profits for shareholders, creating products that are increasingly easy to use. And they may not even realize that pursuit of these mundane goals can cause them to hinder individual choice and control. No company demonstrates this reality better than Microsoft.

The "Software" Company

In the fall of 1995, I saw Bill Gates give a speech in New York City before an audience of a few hundred people. During the question and answer period, Gates was asked what he thought of the increasing dominance of a handful of conglomerates in the communications industry. "This idea of merging companies together," he replied, "is really not a powerful strategy for the information superhighway or anything related to it. . . . Microsoft is not going to make chips, we're not going to own communications networks. We're just going to make software."[7]

Less than a year later, Microsoft was doing much more than just making software. To begin with, it was getting into the business of producing and distributing content. It started MSNBC, an Internet and cable television network, in partnership with NBC; released *Slate*, a leading web magazine edited by veteran journalist Michael Kinsley; and launched the Microsoft Network, an online service with plenty of proprietary content. Soon thereafter, Microsoft initiated Sidewalk, a group of city-based web sites that compete directly with local newspapers, magazines, and broadcast media. And Gates himself bought electronic rights to the world's largest collection of photography, the Bettmann Archive.

Next Microsoft made a series of strategic investments throughout the media, communications, and entertainment industries. It bought WebTV, a leader in the important new TV/Internet hybrid market. It forged financial deals with cable companies, in the hope of getting them to use Microsoft software in the next generation of interactive set-top boxes. It invested in Hollywood properties such as Dreamworks, the movie and music company. It bought or invested substantial sums in all of the leading makers of streaming audio and video technology for the Net. Gates invested heavily in Teledesic, a $10 billion satellite venture. And Microsoft even released an interactive Barney doll.[8]

These developments put Microsoft in position to dominate the future of the interactive media market—whether by WebTV, cable set-top box, PC-based audio and video, satellite, stuffed animal, or some mix of these options. And so, notwithstanding Gates's 1995 pledge, Microsoft soon became much more than a software company.

In fact, Microsoft's ambition was laid bare in an internal memo: "We are challenging . . . newspapers, travel agencies, automobile dealers, entertainment guides, travel guides, Yellow Page directories, magazines and over time many other areas."[9]

It is no wonder, then, that Microsoft came to be perceived as a threat by companies in such diverse industries as computers, media, retail, and banking. They saw a relentless company extending into every imaginable ancillary market to avoid being beaten anywhere. (The effects of this were occasionally even funny: When Microsoft bought the Funk & Wagnall's encyclopedia for a CD-rom edition, the entry on Gates describing him as ruthless morphed into one reporting that "he is known for his personal and corporate contributions to charity and educational organizations."[10])

Certainly, this image of Microsoft as a threat to corporate competitors has been confirmed during its recent antitrust battles. Yet if we dig a bit deeper we can see that individual control is also implicated throughout this conflict. In fact, as much as Microsoft's post–1995 reincarnation was a response to new competition, its lasting effects are being felt just as much by consumers. In other words, the company's attempts to maintain—and extend—its dominance are relevant to us all.

Consider, for example, a particularly revealing statement that Nathan Myrhvold, Microsoft's chief technology officer, made in 1997 in an interview with the *Wall Street Journal*. Microsoft, he said, wants to get a cut of every online transaction—or as he put it, a "vig" (which is slang for the fee a bookie takes on a bet).[11] Microsoft's ultimate goal, Myrhvold seemed to be saying, is not to be a software company, a content provider, *or* a source of electronic commerce—but rather the supreme commercial middleman. This would, needless to say, be something for individual consumers to be concerned about.

Still, a plan to turn the Internet into a toll road and collect a fee as the only gatekeeper might be a bit fanciful, even for Microsoft. Yet the tension between many of the company's practices and the new individual control is already apparent in a number of ways. Even Microsoft's slogan of cyber liberation doesn't seem to hold up. In reality, the company's guiding principle seems to be: Where do you want to go today—within the Microsoft universe?

Microsoft and Free Choice

The tension between Microsoft and individual control begins with the operating system. (An operating system is software that acts as a personal computer's nervous system, coordinating the processor, storage devices, applications, and components such as the keyboard and monitor.) Microsoft's Windows operating system can be found on about 90 percent of all personal computers. Switch on almost any PC in the world, and you will find yourself staring at Microsoft's name and icons. No matter who makes the computer, the fact that Windows is licensed by almost every PC manufacturer means that consumers have little choice when it comes to selecting their most basic software. It means that Microsoft can shape almost everyone's computing experience. And by shaping our computing experience, Microsoft can shape much more.

If the Net is increasingly a lens on life, then the operating system can be thought of as the tint of that lens, for it colors everything we see. Microsoft's current dominance of the operating system market therefore has significance beyond the personal computing industry. As experience is increasingly mediated by the Net, Microsoft has the opportunity to be there each step of the way, guiding and influencing us in subtle and not-so-subtle ways. This would hardly seem to be consistent with an environment in which computer technology will enhance personal control. The control revolution, after all, presumes the ability of individuals to make choices for themselves about information, experience, and resources. If a choice is being made and almost every option involves Microsoft (because of its dominance of the operating system), consumers may well be choosing from a limited range of alternatives.

Microsoft's most ardent supporters, of course, see the situation differently. They interpret the company's dominance not as a challenge to individual choice but as a reflection of it. Microsoft controls nine-tenths of the computer market, these advocates maintain, simply because it has succeeded in getting customers to choose its product.

But how free is that choice? How much does it really reflect what consumers want? Consideration of these questions has to start with the actions of Microsoft itself—particularly with longstanding charges that the company has restricted the ability of consumers to freely choose a competitor's offering.

There is much evidence, for example, that Microsoft has leveraged its position in the software-licensing process to pressure computer manufacturers into using Windows rather than competing operating systems. That, in fact, was what the federal government charged in a 1994 antitrust lawsuit against the company alleging anticompetitive licensing practices. The Justice Department claimed that Microsoft was forcing computer-makers to comply with a "per processor" licensing scheme, which meant that the manufacturers would be charged a fee on every computer processor sold, whether or not those computers actually used Microsoft's operating system.[12]

How did this forced royalty work? The case of IBM is instructive. In the early 1990s, IBM planned to offer PCs to customers with either Windows or a competing IBM operating system called OS/2. If IBM wanted to continue its contract with Microsoft (which it did, given Microsoft's already dominant position), it had to pay a royalty to Microsoft not only on the Windows machines it sold, but on the OS/2 machines as well. This was the equivalent of Microsoft charging computer-makers a tax on every non-Microsoft operating system they sold—a tax that would be remitted straight to Microsoft.

Not only would this do away with any notion of fair competition. It would also make it more costly for individuals to purchase a non-Microsoft operating system, since they would effectively have to pay for Windows too, even though they weren't receiving it. In other words, Microsoft's tax would ultimately be paid by people who were not even its customers. In 1995, the government succeeded in getting Microsoft to sign a consent decree in which it agreed to cease this crooked practice—which was, interestingly, the same type of anticompetitive scheme that had gotten oil baron John D. Rockefeller in trouble with antitrust authorities almost a century earlier.[13]

A few years later, Microsoft came under renewed antitrust scrutiny, as the government and various industry competitors again alleged that it was engaging in anticompetitive licensing practices. In the fall of 1997, the issue was Microsoft's practice of requiring computer-makers that wanted to license its latest operating system, Windows 95, to also license Internet Explorer, its browser. (A browser is software that allows an individual to view content on the world wide web.) The federal government said that such a conditional licensing practice violated the 1995 consent decree. At a press conference, Attorney General Janet Reno asserted that the company was unfairly using its dominance in the PC market to obtain a commanding position on the Internet. Microsoft, in response, claimed that the consent decree gave it the right to develop "integrated products," and that it was simply integrating Windows and Explorer.[14]

Whether or not that was the case with Windows 95, Explorer was certainly integrated into the operating system in its later Windows 98 incarnation. (In public statements, Microsoft employees often cunningly dropped any reference at all to the Explorer browser, speaking instead of the "browsing functionality" of Windows.) When Windows 98 was released in May 1998, the federal government and a coalition of twenty states launched a full-scale antitrust lawsuit against Microsoft, claiming that the integration of the browser and other practices were illegal.

Among the many allegations were these: Microsoft tried to force Netscape to divide up the browser market and keep others out. It threatened the largest computer-maker, Compaq, that a decision to use Netscape's browser instead of Explorer would cause it to lose its right to sell Windows, effectively putting it out of business. It coerced America Online into becoming an exclusive distributor of Explorer by threatening otherwise not to put the AOL icon on the Windows desktop (that is, the opening screen that a user sees when turning on a Windows-based computer). It pressured its own partner, the chip-maker Intel, into not entering the Internet software market. And it licensed software from a small company called Spyglass and then gave away the software, effectively destroying Spyglass's market.[15]

Despite this barrage of claims, some analysts argued that the government lawsuits that began in 1997 represented unwarranted meddling in the so-called browser war between Netscape and Microsoft. But as the U.S. Justice Department's antitrust division rightly recognized, the browser war was not primarily about the browser market itself (after all, both companies were giving browser software away for free on the Internet). Rather, it was about control of ancillary markets—most notably, the market for operating systems.

During the months leading up to Microsoft's decision to integrate Explorer and Windows there was substantial attention in the computer industry to technological developments that might threaten the Windows hegemony. The most celebrated innovation was Java, a web-programming language developed by Sun Microsystems. As a "platform independent" language, Java could be run on any type of operating system—which meant, most importantly, operating systems other than Windows. This flexibility made it popular with programmers, who believed it could become the lingua franca of the web. It also led Java to be praised as a more "democratic" technology than Windows. Java's relatively nonproprietary nature suggested that, although Sun would gain from its proliferation, the company would not have the chokehold on the Net that Microsoft had in the PC arena.

Java, in other words, was thought to be a technology that would preserve individual choice.

At the same time, another technological threat to Windows came from network computers—cheap and fairly unsophisticated terminals that are connected to a central computer and therefore don't need to run on a PC operating system like Windows. Recognizing the importance of these developments, Netscape entered into partnerships with companies like Oracle that were poised to challenge Microsoft in the emerging market for network computers.[16] Netscape also then held a strong majority share in the browser market and Java was on the rise. Many industry analysts believed that if the right partnership was formed—between some combination of Netscape and Java and network computing—Microsoft's competitors might well do away with the need for Windows. A user could get the full benefits of a PC and the Internet using browser software alone.

> Microsoft is making sure that no one undermines its crown jewel: domination of the operating system market.

This is why it makes sense to interpret Microsoft's bid to rule the browser market as a veiled attempt to prevent fair competition from new challengers—by essentially nipping them in the bud.[17] Microsoft was trying to extend its supremacy from operating systems into browser software. But, in a true bait and switch, it was doing so not to exploit the browser market so much as to make sure that no one else could use their position there to undermine its crown jewel: near total domination of the operating system market.

Whether or not the courts agree with this assessment, the important point for the control revolution is this: Efforts by Microsoft—or any company—to use dominant market power to restrict competition are ultimately restrictions on the ability of individuals to make free choices in the marketplace. Intentionally or not, they are an unjustifiable form of resistance to individual control.

This, in fact, is why we have antitrust statutes. With its talismanic focus on competition, it's easy to assume that the purpose of antitrust is to protect a company's competitors. Yet the real goal of this body of law is to protect consumers, particularly their ability to make meaningful, unconstrained choices. The government's antitrust charges against Microsoft, then, were another way of saying that the company was denying individuals a chance to choose freely—and thereby depriving them of control as consumers in the marketplace.

Beyond Preference

Microsoft's defenders, to give them their due again, would object to this assessment and argue that the company's dominance must have something to do with consumer preference. Yet with the exception of die-hard fans of Apple's Macintosh and those sophisticates who gravitate to bare-bones platforms such as Unix and Linux, most people in the market for a personal computer really have no choice but to buy a Windows-based model. The nonprofit Consumer Project on Technology (CPT), based in Washington, D.C., demonstrated this with an ingeniously simple (if not exactly scientific) study of the PC market.

CPT had an intern call twelve leading computer manufacturers—including industry leaders Gateway 2000, Dell, Micron, IBM, Packard Bell, Hewlett Packard, Toshiba, NEC, and Sony—and ask if he could buy a computer with an operating system other than Windows. The answer was clear. No one would sell him a computer without Windows (not even IBM, which made its own operating system, OS/2). The intern then asked whether he could return Windows for a refund. Again, no one would oblige. (IBM's answer was particularly telling: "We don't give refunds on Windows 95, even if you don't want it or don't use it. It comes pre-installed as part of the computer, and you have to pay for it."[18]) The reality, then, is that unless one buys a Macintosh, purchasing a computer without Windows is nearly impossible.[19]

Even those who appear to be choosing Windows freely, then, may be doing so less out of genuine appreciation for the features of the product than out of recognition of its pervasiveness in the industry. This is because of what economists call "network effects," a phenomenon that occurs in certain communications-related markets when individual users of a product gain value from others using the same product. Thanks to network effects, a mildly successful product may get an artificial boost and come to dominate a market, locking in consumers who, but for the prevalence of the dominant product, would prefer a competitor's offering. This means that the competitor's potentially superior product may not get a fair chance in the market.[20]

A classic example is the VHS standard for videocassette recorders. In the battle between VHS and the competing Beta video standard in the 1970s, VHS prevailed because it had the benefit of network effects. Once it got ahead in the market, it was natural for people to choose VHS players over Beta players, even though most experts thought the latter was of better quality. There were more stores renting VHS format tapes, more of your friends would have VHS tapes they could lend you, and so on.

Microsoft's software products have had the benefit of the same network effects. Whether or not they make superior operating systems or word processors or spreadsheets, Microsoft has been competent enough to gain a foothold in the markets for all of these products. With aggressive deal-making and effective marketing, it has produced respectable sales results—and then watched as network effects have helped its products skyrocket in popularity. (According to a late 1998 calculation, the company controlled 89 percent of the market for PC operating systems; 93 percent of the market for office suite software; 66 percent of the market for stand-alone word processors; and 66 percent of the spreadsheet market.[21]) In the world of computing, where standards are crucial, the more people who use these products, the more value they have for each person who uses them. As with the VHS, it will be more convenient for you to be a Windows user if many others are too.

What this means is that Microsoft's overwhelming control of the PC market, and increasingly the Net, cannot be attributed purely to consumer preference. It's not the same as if 80 to 90 percent of people bought a certain brand of sneaker or toothpaste. Purchasers of those products derive no real benefit from the number of others who wear Nike or brush with Colgate. But when it comes to operating systems, many individuals undoubtedly feel that they must select the one that is used by almost everyone else.

Moreover, once they have chosen Microsoft, users are still not able to make unconstrained choices about the tools that will shape their online interactions. That's because Microsoft uses the operating system as a leverage point from which it can extend its reach into nearly every area of digital activity. This is one reason that Microsoft has become the leading provider of software applications such as Word, Excel, PowerPoint, and Outlook. As Microsoft's operating systems—first DOS and now Windows—have become ubiquitous, its application developers have had an obvious advantage over others in terms of getting new products into the market that work well with their operating systems. The goal is to make it irresistibly easy for everyone to compute and to surf the Net the Microsoft way.

Screen Bias

But, as it happens, even that may not be enough. As veteran consumer advocate Ralph Nader has argued, Microsoft's top executives "have gotten themselves into this mindset that, if they don't control everything, if they don't *try* to control everything, they'll control nothing."[22] Perhaps that explains why Microsoft wants to provide more than just the tools that PC users and Net surfers rely on as they navigate the digital world. It wants to be a hub and a destination. It wants individuals to read Microsoft content, watch Microsoft programming, and shop at Microsoft's online stores. That includes web sites that let customers buy a car, purchase real estate, manage their investments, and make airplane reservations. (That's right, Bill Gates's "software" company is now a registered travel agent—and one of the largest at that.)

Microsoft's influence over the experience of the Net user can be seen in a number of its decisions about interface design. In early versions of Explorer, for example, users could customize the Search button so that it would take them directly to the search engine of their choice. (A search engine is a web-based tool that allows a user to find information online.) In later versions of Explorer, though, Microsoft rigged the Search button so that it would only take users to Microsoft-approved search engines.[23] As users of Netscape's browser are no doubt aware, many of Microsoft's web sites, and sites belonging to corporate partners such as Disney and Time Warner, have been programmed so that they can be viewed only with Explorer. Someone using a Netscape browser to visit a Time Warner site called Entertaindom, for example, would receive this message: "Welcome to Warner Bros. You're Almost There! You must first download Microsoft Internet Explorer 4.0."[24]

Microsoft's most ambitious gambit, perhaps, was its bid to capitalize on its built-in audience and become the ultimate information middleman. The first step in this direction—the direction that Microsoft's Nathan Myrhvold boasted about—was its decision to partner with other leading media and retail companies, and to steer users their way via its "channel bar," a feature that a user sees on the first screen when she turns on her Windows computer or returns to the desktop from some other computing activity. The channel bar does away with any pretense about Microsoft being a neutral player providing just the infrastructure for Internet use. It features thirty commercial logos that the user can click to go straight to the web sites of companies including Disney, Time Warner, America Online, CBS, Pointcast, and, not surprisingly, Microsoft itself.

The presence of these links means that Microsoft is directing as many as nine in ten computer users toward the content it wants them to see and the vendors it wants them to patronize. It's a classic case of a dominant access provider giving preferential treatment to its preferred product and discriminating against the products of others. This "screen bias," in fact, has been an important issue at stake in the antitrust scrutiny of Microsoft.[25]

In defense of its channel bar, Microsoft claims that the corporate logos can be removed from the desktop. And it might well add that the channel bar does not actually prevent users from going where they want to go. The truth, though, is that the channel bar's icons represent a shortcut that most users will take (particularly novices and those strapped for time). They're too convenient for most of us to ignore, and Microsoft knows it. Why go to the trouble of charting your own information course when you can follow the well-marked paths laid out before you?

The power of this type of screen bias has actually been well documented. In the 1970s, two airlines, American and United, developed a computer network to give travel agents information about available airline tickets for sale. Though the network listed flights by all airlines, American and United designed it so that their flights would always appear on the top of the first screen. The result was a powerful competitive advantage; one industry study showed that agents chose more than 90 percent of their flights from the first screen and 50 percent of their chosen flights were the top one on the screen. Ultimately, to remedy this unfair competition, the federal government developed anti-bias rules for the airline-ticketing market.[26]

Like American and United, Microsoft has the ability to capitalize on screen bias. But it could build an unfair advantage not just in one field like airline ticketing, but in broad areas such as electronic commerce, banking, news, entertainment, and even social discourse.

Pushing Profit

So is Microsoft trying consciously to limit the experience of Internet users? Or is it just capitalizing on its control of the operating system to make more money for itself and its corporate partners (who will no doubt pay dearly for the traffic they receive)? The answer may be a bit of both, though Microsoft's raison d'etre undoubtedly has more to do with seeking profit than diminishing individual control. In fact, slogans aside, personal autonomy is probably not on Microsoft's radar at all—and that may be the problem. Its executives may not think twice if the quest for revenue requires the company to limit individual control a little bit here and there. Resistance, then, may simply be an unintended casualty of the corporate battle for revenue and market share online.

Microsoft is probably less unique in this regard than it is a very conspicuous example. The company is certainly not alone in using its control of access to steer users to its preferred content. The basic model of online services such as America Online and Prodigy, for example, is not to give people access to the broader Internet, but to offer their own proprietary material. In fact, AOL—which today has a good share of all online users—has arguably influenced user experience online as much as Microsoft or even more. From the start, AOL has prided itself on giving consumers an easy alternative to the chaos of the Net, which generally means content packaged, produced, and pushed at the user by AOL itself. Moreover, AOL was one of the first services to bring commercialism to the Net. When a user signs on to AOL, often the first thing she will see is an advertisement. And she won't be able to do anything until she reads the ad and clicks a button to decline a purchase. It's the equivalent of the phone company requiring you to listen to an ad before getting a dial tone.

In February 1999, customers of Amazon.com were surprised to learn that publishers were paying $10,000 or more to have their books prominently featured on the web site with accolades like "New and Notable" or "Destined for Greatness." There had been no indication on the site that the choices reflected anything other than Amazon's editorial judgment. A few days after the practice was disclosed by the *New York Times*, Amazon said it would disclose to customers when publishers were paying for placement.[27]

> The battle for online revenue and marketshare may cause companies to resist individual control.

Even search engines, those seemingly neutral navigating tools, direct us toward web sites that pay for prominent billing. A few search engines, like Yahoo, Goto.com, and OpenText, have openly tried to sell the top slots of search results to the highest bidder.[28] Others simply sell banner ads that track a user's search, though it is often difficult to distinguish the real search results from the paid links. As one leading search engine, Infoseek, explains in a press release, "When a consumer searches 'fixed rate mortgages,' for example, he or she will arrive at a results page that has been wrapped in real estate related content."[29] Wrapped, that is, in real estate advertisements.

The search engines have also realized the importance of not letting users stray too far. Once their function was to give individuals information and send them on their way. Now, though, companies like Yahoo and Excite are recognizing that in order to make money from advertising and services, they need to keep users at their sites. So they've added news, free email service, chat rooms, stock quotes, and games. And now they are called portal sites rather than search engines. But that's not quite right, either. A portal suggests an entry to somewhere else. These sites, like Microsoft, want to be destinations. They want users to stay, not go.

Indeed, some companies have concluded that the best way to retain the attention of consumers may be to *push* content at them. Traditionally the web works on a pull model, where an individual must take active steps, like clicking on an icon or typing in a web site address, in order to receive information. Push technology companies like Pointcast, though, send a constant stream of content to a computer screen. This might include news updates, data such as stock prices or sports scores, corporate information (in a work environment) and, of course, advertisements. Push technology makes the web less interactive and more like television, in part to do what the tube does best: sell ads.

Push technology was one of the leading Internet stories of 1997, as startups in the field caught the eye of investors and trend-watchers.[30] A year or two later, though, talk about push had cooled. Companies that had billed themselves as masters of push were suddenly recasting their products and writing new business plans. Perhaps this represented a victory of sorts for interactivity and individual control. Yet the first push applications were really transitional technologies, a sign of the increasing convergence between television and the Net that will occur over the next decade.

A hint of that convergence can be found in the rise of TV/Internet hybrids such as WebTV. Introduced in 1996 and purchased by Microsoft a year later, WebTV offers basic Internet access over a television. Relatively cheap and easy to use, it laudably extends the promise of the Net to many who might be discouraged by the cost or complexity of PCs. It even comes with a specially designed remote control for online navigation.

The makers of WebTV concede that they're trying to domesticate the Net. Indeed, the original WebTV sales pitch unabashedly invites the user to be a couch potato: "Perhaps all it takes to enjoy everything the Internet has to offer is a more familiar setting. Since you already have a television, a cozy place to watch, and you already know how to use a remote, you're all set. So just kick back, relax. . . ."

Comforting as this might be to some new users, they could unwittingly be deprived of some of the benefits of real individual choice. More than Microsoft's channel bar, for example, WebTV guides users away from exploring the Net freely and toward preselected, mainstream content. Even the fact that users are supposed to sit eight feet away from the screen and use a remote control is telling. What more could WebTV do to encourage users to be as passive as TV watchers?[31]

All these developments—from screen bias to rigged search engines to push technology—represent a collision between the interests of online companies and the interests of individual users. Despite the seeming inevitability of the control revolution, interactivity and user control will continue to be pitted against the prerogatives of online media companies and advertisers. A world of apparently endless choices may actually be a world of constrained choices, narrowed by actors who seek to gain from our limited scope of options. One notable fact here is that most of these companies support laissez-faire ideals when it comes to government regulation of their own behavior. Yet some of them don't seem to want consumers to have the same far-ranging freedom.

To be sure, Microsoft, AOL, Yahoo, Pointcast, and others can argue that they are presenting consumers with quality choices and, even more, helping them to deal with the confusion of too many options on the Net. At least in the case of Microsoft, though, there is reason to question whether individuals are choosing freely to take advantage of the company's judgment. Rather, they may simply do so because they are using the Microsoft operating system, which is no great sign of free will.

In general, we must ask: Who really determines our online experience? Do we control it or do big gatekeepers like Microsoft and AOL? What are the trade-offs between convenience and real choice? How can we distinguish the illusion of personal control from the real thing? And what can we do to preserve both easy access to the Net and the opportunity to make fully informed decisions about our experience?

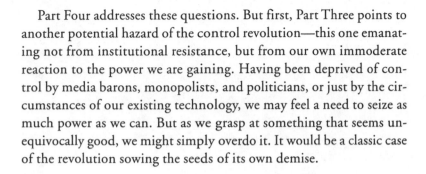

Part Four addresses these questions. But first, Part Three points to another potential hazard of the control revolution—this one emanating not from institutional resistance, but from our own immoderate reaction to the power we are gaining. Having been deprived of control by media barons, monopolists, and politicians, or just by the circumstances of our existing technology, we may feel a need to seize as much power as we can. But as we grasp at something that seems unequivocally good, we might simply overdo it. It would be a classic case of the revolution sowing the seeds of its own demise.

Part Three

Oversteer

"What technology promises is that we can all be control freaks."

Norman Mailer[1]

The American obsession with personal control has always been evident in our love of cars and the freedom they provide us, and the power of the automobile as a symbol of autonomy has not been lost on the computing and Internet industries. Companies like IBM that made the first PCs wanted customers to believe that "the computer was like a car you could control."[2] The symmetry between cars and computers can also be seen in our language. It's no coincidence that a computer that fails to operate is one that has "crashed," or that early aficionados of the Internet described it as an "information superhighway."[3]

Cars and the Net are both technologies that put us in charge. It should therefore not be surprising that a hazard associated with one may also plague the other. The hazard I am thinking of is oversteer. According to Webster's, oversteer is "the tendency of an automobile to steer into a sharper turn than the driver intends sometimes with a thrusting of the rear to the outside." It is what happens when a driver, particularly an inexperienced driver, tries too hard to determine his direction. He exerts so much force that he winds up going farther than intended, with potentially dangerous consequences.

The control revolution presents a similar threat. As we embrace the individual control that the Net makes possible, we may handle our new power so carelessly as to cause unintended outcomes. First, we might overdo our personalization of information and social environments. Second, we could push disintermediation too far, forgetting the value of middlemen in news, commerce, and politics. And finally, we may rely too much on market-based solutions to problems such as protecting privacy. In each case, the dangers of oversteer are not just political, they are personal. Both social justice and individual well-being are at stake.

More than a decade ago, in a book about MIT's renowned Media Lab, Stewart Brand described how "the idea of intense personalization to the user is at the heart of most of the Lab's projects."[4] Brand explained the different personalization projects then under development, including a newspaper that a reader could fully customize. The powerful impact of this form of news delivery was captured in one of Brand's anecdotes about Nicholas Negroponte, the Media Lab's celebrated director. In a speech to executives of the Shell Oil company, Negroponte explained how personalization could affect the relative importance that we attach to different events—events as disparate as a canceled meeting and an attack by Libya's Muammar Gadhafi:

> If last night Mr. Gadhafi had invaded the United States and also last night you had had to cancel this meeting, in my morning newspaper the top headline would be: 'SHELL MEETING CANCELED.' Somewhere down below it would say, 'Gadhafi etc.' In your newspaper Gadhafi might have made the headline, but on your front page someplace it would say, 'Negroponte Presentation Canceled.' To nobody else would the cancellation be news.[5]

In his retelling of this seemingly straightforward story, Brand overlooks some of its most interesting dimensions: What should we think about a media system in which a cancelled meeting can constitute front page "news" while an act of war becomes a footnote, if it is mentioned at all? Should we be concerned about the way in which such customization might backfire and leave us uninformed and unprepared to deal with the world? And what happens to a society in which everyone receives starkly different versions of the news? Is that a state of affairs to which we should aspire? Or is there a point of diminishing returns for personalization of information and social environments—a point at which personalization works against us rather than for us?

Personalization, as we have seen, is one of the major components of the control revolution. The Net is allowing us to exercise more decisionmaking power over the news and information we take in and the environments in which we work, learn, and play. The desire to personalize experience is as basic as almost any human urge toward control. And in an age of proliferating information, personalization becomes all the more necessary. Faced with a deluge of sensory stimuli, we must filter and select just so we can deal with what would otherwise be an overwhelming tidal wave of data. Thus it is not surprising to read an account of a consumer electronics trade show that observes that "much of what was on display were products that help people wall themselves off from the world."[6] And it is no mystery that we do this filtering in a way that caters to our personal interests. Any other way of streamlining the flow of data would seem counterintuitive.

Before considering the larger social implications of personalization, let's look at the impact of this trend on us as individuals.

Personalization of information, of course, predates the widespread use of the Net. People in the same communities have long read different newspapers and magazines, and watched and listened to different news programming. It's hard, though, to imagine any of these conventional media sources responding to a Libyan invasion of the U.S. with anything but an in-depth lead story. Even specialized daily news sources, whether the *Wall Street Journal* or MTV News, would have pieces examining the repercussions of such an attack from their own vantage points. But the dedicated reader of online news who customizes her information intake completely might miss it altogether, especially if the front page of her online newspaper were full of cancelled appointments and other such "breaking news."

> Ignorance and narrow-mindedness are hidden dangers of the control revolution.

The Libyan invasion example is perhaps a bit extreme. One would hope that even the most personalization-obsessed citizen would not be unaware of an event of that scale. But the example frames the issue of personalization in illuminating ways. It shows how oversteer may cause the seemingly innocent tailoring of information to have unintended consequences. What happens when the event in question is important, but not quite as dramatic (or far-fetched) as Libya attacking America? Suppose instead that we're talking about the release of a drug that may help prevent Alzheimer's disease, or a new tax break to subsidize low-income housing in inner cities, or an international treaty to ban landmines. How many of us will program our filters so that the Net brings us this kind of information? Maybe those who work, respectively, on health care, urban policy, or military issues. But will the rest of us, when prompted to indicate our interests, check off preventive medicine, poverty, or world peace? Probably not.

I wonder what important news I missed, for example, when I was trying out Infobeat and Newspage, the news personalization services I described earlier. As I tried each morning to make sense of scores of stories in a few dozen of 2,500 super-specialized categories, surely there were other stories in other categories that I would have wanted to know about, but whose categories I did not select in advance. (And though I assumed that personalizing my news would give me less junk to sift through, Newspage simply gave me more information about fewer topics—so much that I had little time for anything else.)

Ignorance and narrow-mindedness, then, are hidden dangers of the control revolution—hidden because they are self-imposed and, even more, because the Net seems so open and diverse. But in fact the infinite scope of today's information sphere may lead indirectly, if somewhat perversely, to a loss of diverse experience and a flattening of perspective.

Total Filtering

The filtering that is at the heart of personalization is not new. On the contrary, human consciousness has always been characterized by some degree of filtration. We deal with information by sifting and sorting it, using technologies, taxonomies, and other social and cognitive constructs. From an evolutionary perspective, this is an advantageous trait. Without it, we would surely be less well off.

The precision of digital interactive technology, however, raises the impact of filtering to a different level. It makes possible what we might call *total filtering*, the ability to exercise nearly absolute personal control over experience that once was subject only to very approximate control. Basic code features of the Net, notably the power of interactivity and the flexibility that comes with digital storage of information, make such enhanced control possible. Chat rooms, for example, have "ignore" commands, and email programs allow users to automatically delete messages from certain people.[7] Filtering protocols such as the Platform for Internet Content Selection (PICS), which establishes uniform rules for labeling, also make it easier to select or block information on the basis of any criteria.[8]

One danger of PICS is that it can be used by governments, employers, and Internet service providers to restrict the materials that are received by individuals.[9] But we should also be attentive to how individuals themselves use such protocols. While PICS admirably allows any rating or blocking criteria to be used, this change to the code of the Net makes it far easier for individuals to control the information to which they are exposed. By standardizing a protocol for labels, PICS lowers the cost of rating and blocking content. It makes it more likely, then, that information will be rated by any number of organizations (such as a local PTA, a religious organization, or a civil liberties group). Individuals will be able to combine these organizations' ratings with their own content preferences to create highly targeted criteria for what information they receive.

The result is a new level of personal control over experience that—in addition to having implications for freedom of speech—could change our perspective on the world. Certainly, such powerful filtering tools could have many beneficial uses. But it is also not hard to see how we might mishandle them to our own personal disadvantage.

"Each person could design his own communications universe," observes Cass Sunstein, a professor of law and political science. "Each person could see those things that he wanted to see, and only those things. Insulation from unwelcome material would be costless."[10] What we might unwittingly bring about, in other words, is nothing less than the privatization of experience.[11]

Selective Avoidance

It is, as I noted at the outset, impossible to predict exactly how individuals will use the new technology of individual control. Yet a fair amount of psychological research and commonsense logic suggests that we may be inclined to use these tools to narrow our horizons rather than to expand them.

The concept known as selective avoidance, for example, sheds interesting light on how we choose what we see and hear, and with whom we interact. This idea—also known as selective exposure—is part of the theory of cognitive dissonance, which was introduced in the 1950s by social psychologist Leon Festinger.[12] Festinger believed that we all have a desire to produce consistency in our lives. As our minds process different thoughts and stimuli, we want to experience these competing messages in a way that promotes consonance and harmony rather than dissonance. Dissonance, Festinger argued, would be produced when a person had one view of the world in his or her mind but was confronted with information that supported a contradictory view.

"When a dissonance is present," Festinger wrote, "in addition to trying to reduce it, the person will actively avoid situations and information which would likely increase the dissonance."[13] Because of fear of dissonance, he added, psychologists have observed "circumspect behavior with regard to new information even when little or no dissonance is present to start with."[14]

In other words, all of our decisionmaking—about what information to take in, whom to socialize with, and so on—may be shaped by our unconscious desire to resolve contradictions and reduce psychological discomfort. Some psychologists have even concluded that we subconsciously prevent threatening information from entering our minds.[15] Students of Freud will see a connection here to his theory of repression, in which information is selectively forgotten or displaced in order to avoid painful associations.[16] But one need not fully accept Freud's theory, or Festinger's, or any grand claim of psychology to give credence to the idea that, in everyday life, we act in ways that allow us to avoid interactions that make us uncomfortable and information that challenges our beliefs.[17]

Even more fundamental than the idea that we block out challenging or unappealing information, perhaps, is the corresponding notion that we seek out that which we already know about.[18] Quite simply, we are inclined to track the ideas, information, and decisions to which we have already committed. Thus, new car owners read advertisements for the car they have just bought more than they read ads for other cars.[19] And people of Irish descent probably pay more attention to news about the conflict in Northern Ireland than do the non-Irish, whether that news supports or does not support their view. They're more invested in the conflict than others are. Similarly, those who grew up in a family that always watched baseball may well choose to continue watching baseball—not because they're threatened by football or basketball, but because they're just not as used to it.

> Customizing our lives to the hilt would dull our imagination.

The control revolution complicates this tendency to stick with what we know and avoid that which is new or challenging. Cognitive dissonance exists, Festinger says, because "a person does not have complete and perfect control over the information that reaches him and over events that can happen in his environment. . . ."[20] Festinger also believed that, where they could, people would reduce dissonance by changing their environment.[21] But he reasonably assumed that changing one's environment would usually be difficult, if not impossible. Therefore, although people would engage in selective avoidance, they would also learn to achieve consonance by altering their underlying views and behaviors.

This, Festinger said, is a way in which cognitive dissonance can be a positive force in our lives. A smoker, for example, who is confronted with evidence of the health risk of cigarettes may try to ignore that information. But eventually, finding it impossible to disregard the warnings, he may stop smoking. Similarly, a bigot who comes to know and respect someone of a different race may be forced to abandon her prejudicial views. These individuals would be resolving dissonance not by manipulating external stimuli, but by altering the behavior or belief that was causing the tension in the first place.

The control revolution could change the calculus when we are faced with a potentially dissonant situation. Instead of having to reevaluate our actions or views, we would simply use total filtering to shield ourselves from the experience that is provoking us. In doing so, we might succeed in reducing cognitive dissonance. But without such tension, we would likely become self-satisfied and unchallenged, lacking motivation or curiosity. Our preferences would not have the ardor of those that arise among genuinely conflicting voices. Customizing our lives to the hilt, in other words, would dull our senses and our imagination.

The Displacement Factor

A skeptic might note that this is a danger only if we confine ourselves to interacting online. And to be sure, periods of filtered experience will be broken by interludes where we are not totally in control. The same way that typewriters did not make pens irrelevant, and telephones did not eclipse written letters, the opportunity for personalized experience online will not do away with more traditional types of interaction that cannot be easily controlled.

But there is an unmistakable—and critical—displacement occurring. In the twenty-first century, we will have an unprecedented ability to use technology to select the individuals we interact with, the information we receive, the environments we inhabit. The Net will not be something we access only on stationary computer terminals. As we enter an era of "pervasive" or "ubiquitous" computing, the Net will come with us on handheld devices and will be present in many other interfaces around us.[22] It will, as noted earlier, increasingly be a lens through which we experience the world.

Many activities may shift from being offline, public, and somewhat unpredictable to being online, private, and fully controllable: Why walk to a post office when you can send correspondence by email? Why go to a movie theater or even a video rental store when you can order a film online instantly? Why trudge across town to a university auditorium to hear a lecture that you can listen to on the Net? In each of these examples, the displaced experience is one in which the individual would have been exposed to any number of unplanned (and potentially dissonant) encounters with people and information. The new experience, by contrast, occurs in an environment of nearly complete individual control. This is what allows John Perry Barlow to speak of the digital era as "the golden age of narcissism."[23] (Here, it is worth reiterating that "the Net" will include what we today think of as television. The sheer number of hours that people spend watching TV suggests how important the displacement factor is, for those hours will now be subject to a much greater degree of individual control.)

To get a further sense of why the displacement factor matters, consider our information intake when we don't have the benefit of total filtering. For one thing, the avoidance of dissonant material requires a great amount of effort. On a typical day, for example, a middle-aged woman might flip through a magazine, watch a TV program, and listen to people talking as she rides a bus to work. As she does these things, she might be on guard to avoid information that challenges her strongly held beliefs. Suppose she believes that homosexuality is wrong. In a world of limited individual control, she might nonetheless be exposed to an article about adoption by gay couples, a sitcom featuring a lesbian character, or a gay man on the bus discussing a recent date with a new boyfriend. It will be difficult for her to avoid these unfiltered stimuli.[24] As technology proliferates and experience is increasingly mediated and subject to personalization, though, it will be much easier for her to avoid glimpses of gay life that might challenge her views. She can program her software filters to exclude all such information from her news sources (and with telecommuting, she might not even need to take the bus to work and overhear a conversation).

Like this woman, we each encounter much in our daily routines that is unrequested, unexpected, and unavoidable. And like her we will each have an easier time, thanks to the Net and personalization, avoiding whatever it is we want to avoid.

The Feedback Effect

One reason to be especially concerned about this is what we might call the feedback effect. In the course of personalizing my news via Infobeat and Newspage, for example, I chose to receive information about a few dozen subjects, casting aside hundreds of others. The more I learned about those subjects, the more interested in them I became—or so it seemed. In other words, my increased exposure to information about, say, the movie business meant that I knew an increasing amount about that particular industry—its breaking developments, long-term trends, leading players—and so naturally I was curious to know more, to follow along and keep up-to-date. I was experiencing a feedback loop, where a causal relationship was created between the information I received and the information I desired. That desire was then fed back as a request to the source of information, which prompted me with more of what I wanted, and so on.

Informational feedback loops are not just the product of interactive media. The golf enthusiast, for example, who reads golf magazines, watches golf on TV, and purchases home videos that promise to improve his putting may experience the same narrowing phenomenon—where he devotes ever more time to golf, to the exclusion of other interests. But distinct features of the digital, networked world make this narrowing easier, and perhaps more hazardous. So-called "intelligent" software agents, which search the Net for information we request, are designed to give us more of what we asked for in the past and to learn from each request how to select a more perfect match. And the interactivity of the Net speeds up and streamlines the feedback process. Instead of just following movie information, for example, I can delve further into more specific areas of filmmaking—tracking, perhaps, box-office earnings for action films. And I can get more specialized from there.

It's no surprise, then, that there are endless newsgroups, email lists, and other online information sources dedicated to the most specific interests, but you'd be hard pressed to find a Random Interests group, let alone one committed to the General Common Good. Political theorist Jeffrey Abramson observed this splintering phenomenon even in an online community focused narrowly on *Star Trek*: "Not only did this virtual community exclude from the conversations virtually everything outside of the *Star Trek* universe, the group itself splintered into factions that didn't talk to one another, each one dwelling only on its micro-interest."[25]

The pinpointed nature of online information also means that we don blinders when we interact online. As I focus in on the information that I want—or that I think I want—it is nearly impossible for me to get any sense of what I am missing. Unlike the golf enthusiast who catches at least a glimpse of other information as he turns to the golf channel on TV, I can create a hermetically sealed information environment built around nothing but instant gratification.

Another difference between personalization in traditional media and on the Net is the effect of software that tracks a user's actions. Online, every choice I make—whether it's a decision to receive a certain category of news or a one-time click on a particular story—may be recorded by that information provider, aggregated with other personal data, and used to shape my future experience online. For businesses, there is value in this data (see Chapter 15) because they can offer advertisers a more precisely defined audience.

Whether or not a user's choices are tracked, personalization may have unforeseen results. As I make highly specialized content choices I am establishing a set of parameters about my interests that might restrict the diversity of the information I receive in the future. I may reduce opportunities to make new choices about what interests me, while reinforcing choices I have already made. I may, in other words, unintentionally limit myself based on decisions made, perhaps haphazardly, in the past.[26]

At a minimum, this mode of decisionmaking doesn't adequately account for the complexity of personality and taste. At worst, it may cause our preferences to become ever more narrow and specialized, depriving us of a broad perspective when we likely need it most. Ironically, this might leave us in a situation not all that different than if we let a gatekeeper like Microsoft steer us solely toward its preferred content. The answer, as we will see, lies in finding a more balanced approach to control of information and experience.

Chapter 10

A Fraying Net

Ralph Reed never appeared to be someone who would champion unbridled individualism. As executive director of the Christian Coalition, he made his name as one of the nation's most visible spokesmen for traditional "family values." His job was not just to promote the conservative political and religious views of his constituents but to condemn the choices of others—atheists, gays, pro-choice activists. When it came to the Internet, he supported the ill-fated Communications Decency Act and the corresponding view that the medium was overrun with hard-core porn and other "antisocial" material. So I was surprised, when I met Reed at an Internet policy conference, to hear him speak so favorably about what the control revolution meant to his movement.

"As a conservative, I believe the disaggregation of the media environment works in my favor," Reed told me. "I mean, if I have satellite radio and satellite television, the Internet, and web sites, the liberal gatekeepers in New York can't control information anymore."[1]

At first blush, Reed's desire to have Christian conservatives shape a media environment more to their liking than the mainstream media might seem unproblematic. Already the Christian Coalition and a host of other values-oriented organizations encourage their constituents to tune in to certain programs rather than others, to read certain magazines and newspapers. Sometimes these organizations even have their own programs and publications, and this variety of perspectives contributes to a richer communications world.

But again there is the danger of oversteer. The control revolution allows individuals—of all social and political leanings—to insulate themselves in ways that are hazardous not just to themselves but to the general health of communities, including the larger political communities we call states and nations.

Imagined Communities

The control revolution, I have argued, is about who shapes our daily experience—the information we are exposed to, the people with whom we interact. Once individuals had little control over this type of experience. In preindustrial societies, it was dictated largely by agrarian work and traditional communal living. Information generally circulated orally and human interaction was bounded by geography and rigid custom. The advent of the printing press meant a gradual transition beyond experience based solely on observation and the spoken word. By reading, one could "experience" events far away—and do so with far greater selectivity, choosing to read one text instead of another. But few individuals could read and most were subject to the whims of monarchs or nobles.

With the end of feudalism and the beginnings of mercantilism, populations became more mobile and people were able to interact with a wider cohort of individuals. This expanded the range of individual control over experience. Eventually, with the birth of industrialism came the rise of urban centers and mass media such as newspapers and magazines. These developments, along with universal public education, provided individuals with the ongoing common experience that helped foster the nation-state and nationalism—a broader sense of community based in part on shared information and interaction. In many developed nations, radio and television played a central role in defining a national culture, pulling together disparate geographic regions and individuals separated by class, ethnicity, language, and race. These mass media helped to create what cultural anthropologist Benedict Anderson calls "imagined communities," simulacra of what humans had when they lived a more clannish existence.[2]

The point of this thumbnail history is simply that community has always been shaped by common information. As the biblical story of the Tower of Babel warned, civilization depends on shared experience, a collective vocabulary of referents that members can draw upon as they interact and try to remain committed to mutual goals.[3] This is obvious if we think about the building blocks of a culture: language, custom, ritual, myth, religion, law, art, and so on. All rely on facts, ideas, and narratives that describe the shared history and destiny of a band of individuals, that help to answer the question, Who are we? Without shared information, two people cannot come close to answering that question the same way.

In less abstract terms, we are all familiar with the way in which shared experience creates affinity, even if the experience is seemingly trivial. Upon meeting someone new, we are inclined to ask questions that might establish a common link. *Where are you from? Where did you go to school?* And when we speak with members of our own communities, what we talk about is often local—having to do with politics, sports, the economy. Without realizing it, we rely in these conversations on shared information that comes largely from common media sources. *Did you see that article in the paper today? Hear that game on the radio?* Back when communities were smaller and less transient, people relied on direct sensory experience for this community talk. *Have you seen Johnson's farm? It's been untended for weeks!* But as our communities grew in size, this dialogue necessarily began to rely on the shared experience provided by mass media.

As the mass media have replaced the common market, town square, and local saloon or café as primary sources of information, the owners of newspapers and broadcast outlets have gained a huge amount of power over social discourse. Certainly, there has been ample reason to criticize publishers and broadcasters for being excessively commercial and inattentive to the public interest. Yet for all their shortcomings, the mass media also united communities. A generation or two ago, powerful news organs like the *New York Times* and CBS—Ralph Reed's "liberal gatekeepers"—may have fed us a homogenous view of the world, but at least they gave us a common frame of reference as we stood at the proverbial water cooler discussing the day's issues. These mass media provided "a kind of social glue, a common cultural reference point in our polyglot, increasingly multicultural society," according to media critic David Shaw.[4] We could assume that our fellow conversants would be exposed to at least some of the same information. We might even assume some sense of common destiny.

"A No-Longer Shared World"

The personalization ethic presents a direct challenge here, as it is intrinsically opposed to common information. Nicholas Negroponte's example of a customized newspaper that largely ignores a Libyan invasion of the U.S. may be exaggerated. Yet it is telling because it gives us a glimpse of how diverse our information environments are becoming and how this variety may create distance between individuals living within the same geographic communities.

As we know from television today, the more time we spend online, the less time we will have to interact directly with our families, our neighbors, and other community members.[5] But there's a new wrinkle: whereas once the time we spent watching television—or reading a newspaper or magazine—might have provided some sort of shared communal experience (although a poor substitute for face-to-face interaction), personalization provides just the opposite. We may share good times with others online who enjoy the same narrow passions as we do. But the bonds between ourselves and our fellow citizens might become frayed, possibly to the point of breaking.

If members of the Christian Coalition, for example, shunned mainstream news and entertainment, as Reed suggested they should, they might wind up with a pretty skewed view of the world. But, even more, it would be increasingly difficult for Christian Coalition members and nonmembers to get along. The same is true, of course, whether we're talking about the Christian Coalition, the Libertarian Party, or the NAACP. A lack of shared information would deprive different groups of a starting point for common dialogue.

Without even considering the advent of the Net, many social observers already have warned that communal relationships have weakened in recent decades due to hyper-individualism, changing patterns of work and social life, and the decline of local associations such as labor unions, religious groups, PTAs, and fraternal associations. As a result, Americans appear to spend less time with their neighbors and trust one another less. Robert Putnam lent these (much debated) observations an indelible image in the title of an influential essay on the subject, "Bowling Alone."[6] Individual disengagement from public life, he warned, is diminishing social capital—that intangible asset that holds communities together, making our lives secure and meaningful. The excessive customization of news, entertainment, and social interaction would only deplete it further.

> "You can create your own universe. . . . You don't have to deal with people."
>
> —a computer user

Like the threat to individual well-being, this potential for oversteer stems from our ability to exercise a new degree of control over personal experience. In pre-cyber days, as a disciple of one group or another, I may have become ensconced in a world of parochial information and interactions. But it would have been essentially impossible for me to control with perfect precision all the stimuli I received. I would have had to go to great lengths, for example, to avoid a large headline on a newspaper's front page or a breaking news announcement on television. It would, in fact, have been so difficult that I might not even have tried to avoid such exposure. Yet as mediated experience displaces physical contact, total filtering can give us an unprecedented degree of dominion over experience. This means not just getting the news of your choice, but filtering out extraneous facts—as well as interactions with neighbors who are different from you. As historian Theodore Roszak warns, we may retreat into "solipsistic enclaves where the like-minded exchange E-mail with one another and where we choose our own news of a no-longer shared world."[7]

It is not hard to find antisocial sentiments among some of the Net's most dedicated users. As one put it: "You can create your own universe, and you can do whatever you want within that. You don't have to deal with people."[8] Not dealing with people is, of course, at the heart of the problem. Yet the threat to our communities has little to do with misanthropes and hermits. There will, in any culture, inevitably be some who shun social interaction and obligation. The real concern is that these traits may rub off on the rest of us without our really realizing it. As we take advantage of alluring opportunities to participate in idiosyncratic online communities and to expose ourselves only to the information we want, we could unintentionally build barriers between ourselves and those who live among us. The uniqueness of face-to-face contact—with friends, neighbors, teachers, coworkers, fellow citizens—may be forgotten, as may the subtle pleasure of serendipitous encounters.[9]

There are real dangers here to the integrity of communities and perhaps even nations. Local activists might have difficulty competing with online communities for the attention of their neighbors. As a country, we could have difficulty resolving complex questions when we are all experiencing such inconsistent views of what is going on—in essence, such different realities. With fewer shared experiences and information sources, citizens may feel less of a connection with, and less of an obligation toward, one another.

Indeed, as national boundaries become increasingly permeable and irrelevant to networked life, individuals might have less incentive even to identify themselves as citizens of a certain country. Their minimal commitment could be to vote for policies that benefit themselves rather than for those policies that protect the common good.[10] Internationally, even as the global nature of the Net promises to let us shrink the world, compromise between different nations and peoples may be more difficult if we indulge our desire to narrow our horizons. "In the worst case scenario," as E. J. Dionne puts it, "the global village becomes a global Bosnia. . . ."[11]

Ease of Exit

The paradox in this potential for social fragmentation (which, again, is not preordained) is that one of the wondrous qualities of the emerging Net is the way it allows users to break down boundaries, erase distances, and build alliances. The control revolution, as noted earlier, is allowing us to hatch new online communities of individuals located anywhere in the world. These associations have great promise, in terms of their ability to educate and entertain, and to promote learning and even political change. But ultimately they don't have the glue that gives real physical communities their strength.

Put aside for a moment the critiques of online interaction that focus on the deficiencies of the medium: for example, the lack of visual cues—which, after all, might be cured by the emergence of video-based online interaction. Focus instead on the absence of *recourse* online for offensive personal behavior. Sure, there are virtual communities from which you can be "banished" or in which you can be "punished" (beware the virtual flogging). But without any real-world consequences, there is a lack of individual accountability that is inimical to the creation of enduring communal bonds. With online communities, then, we may unintentionally substitute ephemeral ties with others far away for the strong ties of our local environment. (The strength of communities depends on factors that don't have much to do with individual choice or personal control. The happenstance of location, climate, and natural resources, for example, creates dependencies between individuals and groups, and thus creates deep, long-lasting communal bonds. Anthropologists have shown that many cultural practices can be traced to the common ecosystemic needs—in other words, the shared adversity—of people living in close proximity to one another.[12])

It is, in other words, the fluidity of online relationships that makes them weak. It's not just that members can change identities or harass others with impunity. It's that people can exit effortlessly. As Esther Dyson writes in *Release 2.0,* "people who don't like the rules can leave."[13] This is perhaps the most distinct feature of these groups: The ease and convenience of online interaction means that the cost or hassle involved in switching from one to the next is negligible. For me as an individual, this may be a great feature. I can search for the right clique and pick up and move on the moment someone annoys me or I get bored. But it is this very ability to break ties that makes virtual communities ersatz imitators of the real thing. The constant potential for rearrangement may be productive in the short term, but it means that there is little incentive to keep online associations intact.

Some might think that the weakness of these affiliations would ultimately prevent them from posing any real challenge to physical communities. But to the contrary, the ability to meander endlessly from one online gathering to the next, changing habitats on a whim, is precisely the problem. The fact that we can continue exploring and indulging ourselves is part of what makes the virtual life an attractive alternative to getting involved in one's geographic community.

> Our communal conversation could be cut up into an endless number of isolated exchanges.

Few people, we can presume, intend to forge weak bonds online and, in the process, distract themselves from local commitments. But technology always has unintended consequences, and social science research is beginning to show how this may be true for the Internet. Researchers who conducted one of the first longitudinal studies of the Internet's social impact, the HomeNet study, were surprised when their data suggested that Internet use increases feelings of isolation, loneliness, and depression. Contrary to the hypotheses they began with, they observed that regular users communicated less with family members, experienced a decline in their contacts with nearby social acquaintances, and felt more stress. Though a number of scholars have criticized the study's methodology, its results were merely preliminary and the authors were careful to say that more research was necessary.[14] Until more conclusive results are available, what's most important is that we take seriously the hazards outlined in the HomeNet study and attempt to prevent them from becoming worse or taking root in the first place.

Four decades ago, Daniel Lerner wrote an anthropological account of modernization in which he concluded that the introduction of mass media into developing countries is a positive force in part because it allows for a broadening of empathy.[15] Radio, television, and newspapers, in other words, can educate us and get us to care about the plight of fellow community members whose circumstances might otherwise be unknown to us. These media can help to cultivate a sense of common identification and unity. This is, admittedly, quite different than providing a forum for citizen conversation—something the *Times*, CBS, and the rest of our one-to-many electronic media rarely do or ever did. The Net, by contrast, gives each of us some ability to take part in that conversation.

But if personalization is pushed too far, our communal conversation will be cut up into an endless number of isolated exchanges. We would lose our "agora in the media," as communications scholar Ithiel de Sola Pool called it.[16] It would be difficult, maybe even impossible, for citizens to establish common ground, solve problems, and discover compassion for one another.

Chapter 11
Freedom from
Speech

Too much personalization of experience, we have seen, will deprive us of a broad-minded worldview and weaken social and political bonds. It also threatens to undermine the freedom of speech. This form of oversteer may be less obvious than those mentioned above because the control revolution appears to be a boon to free expression. Using a computer and modem, a speaker can make her views known to the world quickly and inexpensively. From the Radio B92 activists to the netizens who took on *Time* magazine's cyberporn story, we have seen how the interactivity of the Net allows individuals to evade censorship and correct fabrications. It all seems like a free speech bonanza—fulfilling both the individual's need for uninhibited self-expression and society's need for robust public debate on matters of pressing importance. Yet, in an environment of nearly absolute individual control over experience, there are ways in which speech may not be free at all.

Consider the effect of the control revolution on a hypothetical speaker whom we'll call Paine. Paine is destitute and his message is unpopular, but he is determined to reach the largest possible audience. In a pre-cyber world, what does Paine do? He goes to places where people gather on a regular basis—urban sidewalks and street corners, marketplaces and town squares—and proceeds to shout slogans, wave signs, and hand out leaflets.

If Paine lived in a totalitarian state, bureaucrats and constables would probably use threats, imprisonment, and violence to cow him into silence. But Paine is fortunate. He lives in a constitutional democracy where the state cannot suppress speech with which it disagrees. Indeed, the classic free speech guarantees outlined by the U.S. Supreme Court over the course of the twentieth century have been secured in cases dealing with weak and unpopular figures like Paine—labor picketers, communists, Jehovah's Witnesses, civil rights protesters, antiwar demonstrators. The Court's "public forum" doctrine, in particular, recognizes that there are spaces in our society in which the speech of an individual cannot lawfully be restricted.[1]

Paine, in other words, is free to be a pain. In the public forums where he speaks, he can protest without fear of government censorship or private retaliation. Even more importantly, he can do so in a way that reaches many of his fellow citizens, even though they may not want to hear him. This delights Paine. He is able to express himself to a known audience without restraint. And though his reluctant listeners wouldn't be quick to admit it, many of them recognize the value of having Paine around. Occasionally he says something that makes them think twice. Or at least he reminds them that their view is not the only one out there.

Now consider Paine's plight in a world of the Internet and total filtering. Recognizing that he must follow his former audience to where

> The control revolution is changing the ground rules of free speech, allowing the dissenter's voice to be excluded effortlessly.

they now congregate, Paine goes to a local community center and uses a public computer terminal to go online. With a little help, he even creates a web site that contains all his best rants. But he is unable to get anyone to visit his site, because he doesn't have the means to advertise it online or offline. (He doesn't, for example, run commercials on TV that say www.Paine.com.)

So Paine tries to use email to get attention. He gathers up email addresses and sends out his rants, but quickly finds that this doesn't work either. Junk email and other forms of data smog have left Paine's potential audience resentful of unsolicited messages. They ask him not to send them any more email or they set their email programs to block all messages from him. Or even worse, many of them have decided already to accept messages only from preapproved individuals; they don't even need to know that he exists in order to keep him away. The same happens when Paine joins various online discussions. The power of total filtering means that his voice can be excluded effortlessly.

To see what this does to free speech and social discourse, consider what would happen if we reversed Paine's story. What if Paine's reluctant listeners could take the Net's personalization of experience and use it in the physical world? As they prepared to walk through a crowded public square, they could program a filter to erase Paine or anyone else. They would not have to hear the civil rights marcher, take a leaflet from the striking worker, or see the unwashed homeless person. Their world would be cleansed of all interactions save those that they explicitly chose. The cumulative effect on Paine would be little different than if someone placed a glass bubble around him, or if he were removed from the public square and permitted to protest only out in an empty field at the outskirts of town.

Confronting the Problem

Paine's predicament shows how the control revolution is changing the ground rules of free speech and, consequently, of civil society. Even as new technology gives individuals the ability to speak without fear of institutional censorship, it gives all of us a new ability to avoid speech we don't want to hear. The result, in the aggregate, is that the speech of certain individuals—especially marginal speakers—may well be lost in cyberspace.

This problem is not confined to the Net. Prior to the rise of new media, traditional public forums have been dwindling in number and importance, leaving speakers like Paine without an effective soapbox. Suburbanization has allowed the middle and upper classes to isolate themselves, avoiding not just the blight of the inner city but the need to confront those who are different. The conversion of formerly public spaces, such as marketplaces and parks, into private enclaves has also contributed to the sanitizing of public experience. As Yale law professor Owen Fiss has shown, the unconstrained cacophony of the street corner has given way to the restrictive rules of the shopping mall and the privately managed commons. Moreover, where true public forums continue to exist, courts have sometimes narrowed the scope of free speech that is allowed there.[2]

What may be most distressing about total filtering, then, is the way it could solidify a trend toward the elimination of spaces where citizens can confront and engage one another. Of course, it's always been the case that some speakers have a hard time getting noticed, and this is not always a lamentable fact. The difference, though, is one of opportunity.

The American system of freedom of expression includes a kind of unspoken compromise between the unpopular speaker and the reluctant listener. Though we may take it for granted, there is a careful balance of power that gives the speaker a minimal opportunity to be heard and the listener the freedom to move on after fulfilling her equally minimal, even subconscious, obligation to acknowledge the speaker. We presume, then, that a speaker will get at least one bite at the apple—one chance to confront passersby and capture their attention before they avert their eyes and continue on their way.[3]

On its face, this opportunity for confrontation may seem more of a luxury than a necessity. So long as people are not censored by government and can choose to hear what they want, some might say, then free expression will flourish. But this ignores the reality of how speech works. Governing regimes that have sought to muzzle critics have often simply deprived those speakers of an audience. They don't actually need to prevent a rabble-rouser from talking. They can just refuse to give him a license to picket or parade. Or they can threaten to arrest him for loitering or disturbing the peace. Or if he gets really wily, they can remove him to some Siberia where he is sure not to be heard.[4]

The opportunity for one citizen to confront another is important not just for the sake of the speaker. It is fundamental to the idea that in a democracy "debate on public issues should be uninhibited, robust, and wide-open."[5] From the collective decisionmaking of ancient Athens to the town meetings of colonial New England, open dialogue between citizens has been recognized as a key component of a democratic society, allowing unpopular ideas to compete with orthodoxies.[6] It ensures that citizens will hear the complaints of society's most aggrieved, rather than automatically filtering them out. On a day-to-day basis, this may not seem to matter much. But at the crucial moments that shape politics—a civil rights protest, a labor strike, door-to-door canvassing for a reform candidate—nothing may be more important than a person's ability to temporarily dislodge fellow community members from their worlds of individualized control.

Surely there are—and must be—many private environments in which we can shield ourselves completely from unwanted stimuli. Indeed, the overwhelming majority of our time should be spent in such environments. But our constitutional system also presumes the existence of public spaces in which we cannot fully privatize experience.

"Outside the home," the Supreme Court has said, "the balance between the offensive speaker and the unwilling audience may sometimes tip in favor of the speaker, requiring the offended listener to turn away."[7] Or as the Court plainly put it on another occasion, "we are often 'captives' outside the sanctuary of the home and subject to objectionable speech."[8]

In the digital context, things are different. Though the Net empowers us as speakers, it empowers us as listeners even more. We need never be "captives" subject to speech we don't want to hear. There is, in other words, no reason to believe that the speaker will get his one bite at the apple.

This leads to an interesting "If a tree falls in the forest . . ." question: If a figure like Paine speaks and everyone sets their filters so that they don't hear him, is he speaking freely at all? Certainly, such a situation does not help us, as a society, to achieve the broad democratic aims of free speech. Yet many First Amendment traditionalists would still say that Paine is freely exercising his speech rights, and their reasoning would go as follows: Paine is not being prevented from speaking by the government. Rather, he is simply bringing an unpopular offering to the marketplace of ideas. If people found what he had to say worthy, then they would not block his messages. Indeed, if Paine's ideas were compelling enough, they might even come to be seen as more valid than the ideas he was criticizing.

The problem with this argument is that it assumes that Paine has had a fair shot in making his views known to the public. Yet we have already seen how total filtering may deprive Paine of even the minimal opportunity to engage others that he had in the traditional public forum.

The New Market for Speech

So are Paine's ideas really being rejected by the marketplace of ideas? Or is the marketplace not open to him in the first place? The marketplace of ideas is, of course, a metaphor. Yet as conceived by the famous jurist Oliver Wendell Holmes, it presumes an accessible, merit-based forum in which an idea can be aired.[9] The fate of that idea is then supposed to turn on its quality and veracity, not on its initial popularity or the resources behind it. Our free-speech tradition, in fact, presumes that one reason we don't censor "bad speech" is because it will ultimately be trumped by "good speech" in the marketplace of ideas. Of course, there have always been different levels of access to this marketplace depending on a speaker's status and wealth. But the control revolution presents the possibility of a much more cutthroat market for speech. When potential listeners can effortlessly screen out unwanted views, the ability of speakers to have their expression heard will depend increasingly on their ability to penetrate barriers of exclusion.

The odd thing is that the interactivity of the Net appears to remove such barriers, yet filtering technology allows them to be erected with such ease. As Bill Gates says, "We'll exercise more control over who can interrupt us, who can get to our in-box than we have today where people can ring your phone or ring your beeper. . . . Controlled access will be a big theme and software can take care of that."[10]

> The new market for speech combines the traditional marketplace of ideas with a new market for a very precious resource: attention.

Similarly, one of the great promises of the digital age has been the idea that it will be inexpensive to speak and reach a wide audience.[11] Yet in the new market for speech, speakers may well have to pay for an audience. That may be the only way that they can get people's attention. Marketers who want individuals to read their commercial pitches seem increasingly willing to pay them to do so.[12] And when it comes to speech directed at politicians, our campaign finance system has already shown how costly it can be for a citizen to be heard.[13] Is there reason to think the situation will be any different for digital speech between one citizen and another?

Again, listen to Gates: "If a stranger . . . wants to send you [electronic] mail, [he'll] have to put up a certain amount of money in order to get you to read it because your time is the valuable resource."[14]

The new market for speech can thus be seen as the combination of the traditional marketplace of ideas with a new market for a very precious resource: attention. Information overload is driving citizens to narrow their speech environments by personalizing the information they take in. The interactivity of the Net, in other words, is creating an oversupply of speech, particularly commonplace citizen speech, relative to its demand. Demand is low because listeners' attention is stretched thin. People already have too much information to process without having to read an email from some random citizen who wants to talk about why his taxes are too high. They have no time for it, and so they filter it out.

This market for speech has tangible consequences. It may, as noted, undermine the existing balance of power between the unpopular speaker and reluctant listener. But it also affects the ability of the average person—or the average company—to speak and to be recognized.

Different versions of the pay-to-be-heard model described by Gates can, in fact, already be found on the web. Earlier I mentioned how some search engines have begun to auction search results to the highest bidder. That means that the speaker who pays most in a certain category—whether it's dermatology, dog food, or democracy—comes up in the top slot when a user does a search in that area. And most of the other search engines simply place search results amid targeted ads that track the user's search. Type "bookstore," for example, and a prominent paid link to an online bookstore—a so-called banner ad—will appear sooner than a link to the web site of any small bookstore. Gateways like America Online charge steep fees for a prominent spot on their sites.[15] In all these situations, major commercial speakers are using their financial resources to get recognized, while the voices of smaller companies, nonprofits, and individuals get jumbled in the mix. Their speech, like Paine's, may effectively be silenced.

If the new market for speech complicates our ability to speak, at least it would appear to benefit us as listeners. Yet the benefits are not so clear when we consider the multitude of perspectives unknown to us simply because they are espoused by underfunded speakers. The market for speech, in other words, further limits our opportunity to have accidental encounters that may help us in ways we didn't anticipate. We have seen the importance of such encounters already, in terms of staying informed and maintaining community. They are also central, though, to democracy. We cannot make informed decisions about social and political issues unless we are exposed to a wide range of views. This includes some speech that we might not initially want to know about. And it includes some that we might want to know about, but which might not be available to us because of the speaker's lack of resources.

For example, suppose Paine wants to inform his fellow citizens about the racially discriminatory employment practices of a popular retail store. If the store is located in physical space, Paine can stand on the sidewalk outside the store and let all the patrons know why he thinks the proprietors are biased. What if the store only exists on the web, though? How does Paine reach the store's patrons? The easiest way would be to purchase a banner advertisement on the store's web site, but that would be prohibitively expensive and the store, furthermore, could refuse to sell it to him. Entering an online chat forum run by the store probably wouldn't work, because as soon as he started talking they could toss him out (with impunity, since the site would not be a public forum). Paine's only real option would be to set up an independent protest site on the web.[16] Even if he could afford this, he would likely fail to attract an audience—or at least the specific audience of patrons he wanted to reach. His efforts would be for naught and the store's patrons would be deprived of an important message that they otherwise would not know about. The store's bigoted proprietor would be the only one coming out ahead.

With the new market for speech, then, the foes of unfettered talk may have it easy. Silencing dissenters would seem to require little action at all. Instead of ordering the city constable to arrest the protester or having him ejected by private security guards, the owner of a web site or online chat forum could simply press a button and—poof!—the dissenter would be erased. (Speaking about the ability to use the Net for human rights activism, organizer Bill Batson cites Frederick Douglass and says, "Power concedes nothing without a demand. But making that demand with a device that can be turned off . . . is not very compelling."[17])

Paine's problem—and ours—is that there are no real public forums online, no communal areas in which individuals are occasionally subjected (as a matter of law) to speech that they don't want to hear. Admittedly this may not seem at first like a problem. It may, in fact, seem to be a wonderful new benefit of the control revolution, the ability to maintain one's solace and sanity by filtering out all the noise of an information society. But there is something different about the individual control at work here.

So far, I have been talking about the consequences of people controlling their own lives, not controlling the lives of others. I've argued that sometimes too much personal control may cause us to miss important experiences or to lose the benefit of certain ties to others. But here we have an instance where the freedom of some individuals (to filter out speech) may begin to interfere with the freedom of others (to speak to fellow citizens). In other words, we are seeing how the control revolution's shift of power could benefit some individuals more than others—or even some individuals at the expense of others.

Having discussed three types of oversteer related to personalization, the next three chapters consider ways in which too much disintermediation can work against us.

Chapter 12

The Drudge
Factor

Matt Drudge is right where he wants to be: front and center, shaking things up, making everyone squirm a bit. It is late spring of 1998 at Harvard University's cavernous Memorial Hall, site of the Second International Harvard Conference on Internet and Society. On the stage, the distinguished law professor Arthur Miller is leading Drudge and half a dozen other panelists in a Socratic dialogue on news in the digital age. The panel includes prominent editors and writers, an esteemed professor of media studies, a leading cyber lawyer, and a well-known Washington pundit. For Drudge, the infamous Internet columnist who distributes his near-daily *Drudge Report* online, the formal setting is just right. Drudge's routine, after all, is about thumbing his nose at the establishment.

Miller spins out one of his intricate fictional scenarios for consideration by the group. It's about an Internet hack, just like Drudge, who publishes an online journal that is full of controversial claims. Miller grills the panel with point-blank questions: What are the writer's responsibilities to the public and to the truth? How should traditional media organizations respond to him? Should he be held to the same legal standards as other publishers? Is he a threat to the republic—or a savior?

Though the questions are hypothetical, it is clear to the panelists and the few hundred audience members that it is Drudge who is on the spot. Drudge's name has been synonymous with concerns about the accuracy of information on the Net ever since his erroneous August 1997 report that Republican operatives had court records showing that Sidney Blumenthal, an aide to President Clinton, had beaten his wife—a claim Drudge retracted a day later. (Blumenthal nonetheless sued Drudge for defamation.)[1]

As Miller presses his questions, most of the panelists agree that the ability of everyone to be a publisher is a positive development for free speech, democracy, and humanity generally. But many express concern that, without a sense of limits, this new individual power may cause our information universe to become even more plagued by half-truths and lies than it already is.

"We're going to see reputations destroyed, stocks plunge, people lose money, and then maybe months later, years later, it will be 'oops, it wasn't true,'" says panelist Norman Ornstein of the American Enterprise Institute. "There's going to be a whole lot of collateral damage."

Drudge, his brow furrowed in mock-detective concentration, does little to convince anyone otherwise.

"We're entering the era of the citizen press where everyone's going to be a reporter and has the right to report, not just 'legitimate' news organizations," he says bluntly.

With no experience in journalism prior to launching the Drudge Report, Matt Drudge certainly has the right to distinguish himself from the professional media. Working from what he describes as a "a moldy apartment just off Hollywood Boulevard," he eschews the traditional conventions of journalism—original reporting, double sourcing, fact-checking—in favor of a vacuum cleaner mode in which all tidbits are collected and the most sensational are spewed out. In the spirit of infamous tattlers like Walter Winchell, he prides himself on having made his name by disrupting the accepted workings and assumptions of the major media organizations. He scoops their breaking news and runs the stories that they don't think are fit to print.

"The reason I'm succeeding," Drudge tells the audience, "is I reject the corporate notion of news—controlling news cycles, embargoing things, killing stories."

Drudge's first real coup, in fact, was his exclusive report that

Newsweek had decided not to publish a certain story suggesting that President Clinton had had a sexual affair with a former White House intern, and that independent counsel Kenneth Starr was investigating a possible cover-up of the relationship. Here's how it began:

01/17/98 21:32:02 PST—NEWSWEEK KILLS STORY ON WHITE HOUSE INTERN XXXXX BLOCKBUSTER REPORT: 23-YEAR-OLD, FORMER WHITE HOUSE INTERN, SEX RELATIONSHIP WITH PRESIDENT

The implications of Drudge's report, as we now know, were devastating. Within hours, his account was mentioned on one of the Sunday morning television news programs by conservative commentator William Kristol. "The story in Washington this morning," Kristol said, "is that *Newsweek* magazine was going to go with a big story based on tape-recorded conversations, which [involve] a woman who was a summer intern at the White House." Immediately, former Clinton advisor George Stephanopolous dismissed Kristol's reference: "And Bill, where did it come from—the *Drudge Report?*"[2]

From there, the major news outlets were all over the story, and the rest—denial, scandal, impeachment, acquittal—is history.[3] Technically, it all started with a report typed and emailed from the home computer of an irksome Internet gadfly. Of course, even without Drudge, news of Clinton's affair with Monica Lewinsky would eventually have broken. But the facts themselves might have unfolded differently. *Newsweek* reportedly held the story at Starr's request because he wanted Lewinsky to wear a wire and tape incriminating conversations with Vernon Jordan and perhaps even Clinton. Once Drudge hit Send, that plan was dashed. The actions of one individual online, in other words, may well have shaped the political scandal of the decade (if not the century).

Perhaps no form of disintermediation is more starkly apparent today than the removal of layers of news middlemen that the Net allows—and that Matt Drudge personifies. Where once there were reporters, writers, editors, fact-checkers, production staff, publishers, libel lawyers, and large media owners, now a worldwide dispatch may be the end result of a quirky thought and a bit of tapping at a keyboard in one's bedroom.

The emergence of small online publishers has confounded the journalistic establishment. Should these neophytes be treated as colleagues— able to claim the legal and professional privileges of traditional journalists, and eligible for the industry's top honors and prizes?

The shift in power dynamics that is causing this bewilderment has many salutary results, as we have seen. Major news outlets will be held more accountable—like *Time* magazine in the case of its flawed cyberporn report—and upstarts may be able to find an audience and diversify the media environment. But as with excessive personalization of experience, there is a downside, a point at which the advantages of news disintermediation are outweighed by its disadvantages.

No Need to Check

To begin with, the accuracy and integrity of information may be uncertain when it comes from sources that don't have editors and fact-checkers, and major reputations at stake. This is mostly a function of resources, as it takes time and expertise to do careful reporting and checking of facts.

I remember asking Drudge at the Harvard conference whether, with his increasing prominence, he had any plans to hire a staff to help him put out the *Drudge Report*. He scoffed at me and said no, as if I just didn't get it.

Now, I think I do get it. The Drudge factor—that is, the extreme disintermediation of our information environment—means that responsibility for determining truth rests as much with those who consume information as with those who produce it. This is an archetypal example of the control revolution, for it represents a clear transfer in power over one of our most important social functions: who distinguishes fact from fiction and ultimately determines what is true. Increasingly, we bear that burden.

For figures like Drudge, the beauty of this is that they can appeal directly to our innate desire for total control. *Who are we to make authoritative claims about truth? You decide!* It means Drudge never has to say he's sorry if what he publishes is wrong. In a sense it can't be wrong, because he simply reports what he hears in offhand conversations with reporters, editors, and sources.[4] *A scandalous newspaper story is in development; a breathtaking magazine article is being killed. Rumor has it that it says X (or maybe Y or Z).* Drudge needn't have any confidence that the underlying claim is right, since he's merely relaying what someone else is researching or writing. Yet these individuals, who are closest to the story, may also have expressly decided that it cannot yet be reliably reported.[5]

> Increasingly, responsibility for determining truth rests as much with those who consume information as with those who produce it.

No Time to Check

The Net's many-to-many interactivity means that there will be a steady increase in available information and, correspondingly, an increase in the number of mendacious smears, dangerous distortions, and wacky conspiracy theories that float around. The split-second speed at which digital material can be distributed anywhere allows these half-truths and lies to spread instantly and with little time for verification. This increases the pressure on all content providers, including the most venerable journalists and media organizations, to cut corners and lower standards in order to get the story first—or at least not to lag far behind.

Here is an example, then, of how increasing individual control not only can make the standards and performance of middlemen less relevant, but can actually subvert them. The same forces that may induce web publishers to play fast and loose with the facts are affecting our most prominent media companies. These organizations are, increasingly, disintermediating themselves—migrating to more rushed and unfiltered news coverage in order to keep up with the quick pace of the Net.

"Feed the beast. File. Now. Now. Now!" That's how one newspaper reporter describes the pressures of "twenty-four-hour cyber-cycle" journalism.[6] On television, we see a rise in live "spot" news coverage and talk programs where nonreporter "experts" speculate about events as they unfold. Magazines and newspapers similarly rush stories into print—or, even better, onto their web sites—only to retract them hours later.

In the early days of the Clinton-Lewinsky scandal, for example, the *Dallas Morning News* placed an article on its web site stating that a Secret Service agent was prepared to testify that he had seen Lewinsky in a "compromising situation" with the president. Hours later, the story had been removed with a statement from the *Morning News* that it was wrong. The next morning, however, the print version of the newspaper carried the story with modifications, and it was retracted altogether on the following day.[7] Shortly thereafter, the *Wall Street Journal* similarly published a breaking story on its web site stating that Clinton's personal steward had told a grand jury that he had seen Clinton and Lewinsky alone together. The report was immediately criticized by the steward's lawyer as false and irresponsible. The next day the *Journal* published a revised version of the story in its print edition.[8]

In a sense, these newspapers may have been using the web to test the validity of a controversial story before committing to it in print. By its nature, an article published on the web is always in "release 1.0," to use the lingo of software developers. The fact that it can be modified so easily and imperceptibly prevents it from ever really being a final draft. Instead of running a correction, an author can simply change the text as it appears on the web. The problem is that this ability to make seamless corrections after the fact can create a journalistic atmosphere in which sloppiness and inaccuracy are tolerated. With the pressure to be first to get the story out, the "publish now, edit later" school of journalism may become widespread.

No Way to Check

Another danger is that due to the novelty of the Net we may not yet have developed a sophisticated eye for judging the veracity of online information. The ability to make such communications anonymous or pseudonymous, or to manipulate dates or other identifying information, means that someone's random conjecture or joke can easily be mistaken for an assertion of fact—whether it is intended to be taken that way or not. Digital signature technology will likely allow us to authenticate who really sent an email (such as one that says it's from president@whitehouse.gov). But beyond the simple question of whether the sender is really who she says she is, critical faculties for the evaluation of information online have yet to be developed widely.

As novelist and critic Umberto Eco says, "After years of practice, I can walk into a bookstore and understand its layout in a few seconds. I can glance at the spine of a book and make a good guess at its content from a number of signs. If I see the words Harvard University Press, I know it's probably not going to be a cheap romance. I go onto the Net and I don't have those skills."[9]

This type of complaint, which is commonly heard today, should subside in years to come. Already, information brands are being established online—some from the world of old media, some intrinsic to new media. And Internet users are developing experience that allows them to distinguish the cheap from the refined, the dubious from the accurate. By the time photos of a mauled Princess Diana in her crashed Mercedes showed up on the web in September 1997, for example, most Internet users were quick to question their authenticity.[10]

Still, because of the fast-morphing nature of information online—its ability to be there one day and gone (or changed) the next—the likelihood of being bamboozled may always be greater online than in traditional media. Who could have known that a commencement speech circulated widely on the Internet and purportedly written by novelist Kurt Vonnegut would turn out to be a newspaper column written by a columnist for the *Chicago Tribune*?[11] Even the existence of fluid hyperlinks between web pages poses difficulties. While it allows readers to gain depth and perspective on a story, it also means that a few clicks may take them from an established web site to an offbeat site to one that specializes in delusional speculation—often with little indication that they've switched information sources.

In this environment, it will also be more difficult to distinguish editorial content from advertising, or to tell whether there's really just an accomplished charlatan behind a web site that says "published by Big-Time Press," "classes offered by Big-Time University," or "manufactured by Big-Time Company." What's at stake is not just our news intake, but all forms of fact-gathering—by academic researchers, businesses, human rights activists, and so on.[12]

We Have Met the Enemy . . .

Having seen the dangers of oversteer in this context, we need to be clear about who is responsible for these hazards. It is, of course, tempting—and easy—to point a finger reflexively at the Drudges of the world. Yet simply blaming provocateurs and firebrands will do us little good. It could, in fact, cause us to miss the real challenge at hand. In a culture of disintermediated news, threats to truth have as much to do with consumers of information as with its producers. That everyone can be a publisher may be disconcerting, but it is not itself a great danger.[13] Rather it is this fact combined with the illusion that we can all be our own editors and fact-checkers that gets us into trouble.

Paradoxically, then, a culprit here is the celebrated access to information that the Net allows. Before the Net's emergence, anyone could print up a leaflet full of conspiratorial nonsense. But their distribution methods were limited. Geography and cost were strong barriers that prevented false information from getting far. These factors acted, in a sense, as filters that prevented individuals from consuming unreliable information. The Net, by contrast, makes it easy for a few prevaricating crackpots to spread lies—and for us to read them, naïvely believe them, and irresponsibly pass them on.

Yet misinformation is only really dangerous when there is both an unreliable source *and* a credulous audience.[14] As the amount of questionable material increases, then, we need to be ever more cautious and skeptical. Indeed, the control revolution is blurring the distinction between news professionals and audiences, forcing us all to deal with the same predicaments. The common challenge is one of exercising self-restraint to prevent the spread of inaccuracies. On the one hand, that means not being the originators of flawed information (though obviously, few of us intend to do that). On the other hand, it means exercising caution as information consumers. Do we blindly believe what we read? Do we weigh the accuracy of different content providers? Do we pass along, without warning, information that we know comes from dubious sources?

Even as journalists and citizens become seemingly interchangeable, though, a fallacy of the new individual control is the idea that each of us can, or should, judge the accuracy of information on our own. We shouldn't be our own editors. Rather, we should *choose* our own editors, and do so carefully. As I'll discuss further in Chapter 18, what we need is not less mediation, but more effective mediation.

This is, admittedly, a tall order. During the weeks following Drudge's challenge to the establishment at Harvard, one journalistic scandal after another broke at top media organizations—as if to prove Drudge's claim that he is no worse than the old guard. A writer at the *New Republic* was fired for littering dozens of stories with fabrications that, in retrospect, seemed so obviously false that even a high school editor should have caught them (and he was initially exposed in an online publication). Two well-liked *Boston Globe* columnists were dismissed for journalistic misconduct. And more heads rolled when CNN and *Time* magazine retracted a much-hyped joint project in which they had accused the U.S. military of using nerve gas on American defectors during the Vietnam War.

Traditional news outlets, then, have no monopoly on virtue when it comes to the reliability of information. There is a difference, however, between their lapses and that of a hack who publishes from his home via email. It's that no matter how much people complain about the hack, he can't be fired.

As Drudge himself once said in response to criticism from the White House press secretary, "What is Mike McCurry going to do, call my boss?"[15]

Chapter 13

Shopper's Heaven?

Like many people, my first purchase on the Internet was a book. So was my second purchase. And my third. Soon, I was buying all my books this way. I was becoming a model digital-age consumer, using the Internet to bypass local bookstores (those quintessential middlemen) and take greater control of my commercial transactions.

Bill Gates calls this "shopper's Heaven"—and initially, I was inclined to agree.[1] Online bookstores are informative and have nearly everything in stock. Most of all, they save time. In seconds, with a click of a mouse, I can have a book sent to me (or a friend) for arrival the next day. They're cheap, too. In fact, sometimes I browse for titles in neighborhood bookstores—who doesn't love the tactile feel of a book?—and then go home and order them on the Net to get the steep discount. The success of outlets like Amazon.com suggests I am hardly unusual.

There's something ironic about this: Here we are stocking up on pulp using the cutting-edge medium that was supposed to bring about the demise of the printed word. The rise of online book-buying, then, is another reminder of how unexpected the impact of technology can be. It's unexpected, though, not just because computer networks have yet to produce the paperless society, but also because the social and economic impact of electronic commerce may be quite different from what we expect. As with news and politics, we should recognize that too much disintermediation in commerce can backfire.

The Cybernetic Wal-Mart Effect

Middlemen make up a huge sector of our economy and a good percentage of the modern workforce.[2] In the near term, then, the move toward electronic commerce could substantially disrupt conditions of labor and productivity. Times will be hard for middlemen who cannot adapt to a new mode of commerce. Businesses will fold; companies will lay off employees.

In product markets, the effect of disintermediation is being felt already in the area of digital goods such as music and software. The retail chain Egghead Software, for example, closed the last of 200 stores in February 1998, laying off some 800 people and moving its remaining 200 employees to Portland, Oregon, to sell software via the Internet.[3] In the market for services, information providers are going to be the ones affected most: stockbrokers like my Uncle Max who have to compete with online trading services, or travel agents who now have to go up against Microsoft. The once-exclusive housing databases used by real estate brokers, for example, are now becoming available to the public, depriving those brokers of much of their stock in trade. Even professionals such as lawyers may find that their clients are using the Net and other new technologies to avoid having to pay top dollar for legal advice.[4]

Richard Sclove of the Loka Institute, an organization that studies the impact of technology, predicts that as commerce moves to the Net we will experience a "cybernetic Wal-Mart effect." Many local businesses will be unable to compete with their giant new online competitors, just as small family stores were unable to keep up when huge chains like Wal-Mart or Kmart moved into their area. But Sclove predicts things will be worse this time:

> Online, you're not just competing with the Wal-Mart on the outskirts of your town. You're competing with the full global marketplace. Wal-Marts basically were a threat to Mom-and-Pop retail shops. Online commerce can spread out into virtually every sector of the economy. So local service providers—lawyers, stock brokers,

insurance agents, travel agents, all those kind of folks who formerly were competing with each other in local economies—are suddenly competing nationally or even globally.[5]

The repercussions could become evident not just in economic statistics, but in our everyday lives. As consumers, we may be deprived of the presence of many cherished stores—such as bookstores. If they disappeared, we would lose not just a convenient point of commercial exchange but a communal gathering space that is a vital part of our public sphere. The best bookstores, after all, are literary dens where you find not just an intriguing selection of titles but an oasis of culture and calm.

In recent years, there has been a sharp debate among book lovers regarding the rapid growth of "superstores" such as Borders or Barnes and Noble, which appear to be driving smaller, quirkier bookstores out of business. It may turn out, though, that the real threat to small bookstores will come not from the superstores but from online booksellers and from consumers (like me) who have flocked to them.[6] Indeed, since they are both intermediaries, small and large bookstores alike may struggle in years to come. As electronic commerce grows and individuals have the opportunity to do business directly with those who produce books and other goods and services, many traditional middlemen will lose revenue and see their profit margins dwindle. Only those who can adjust and provide unique value to customers will survive.

Lower prices, wider selection, and greater convenience are certainly worth two cheers. Bibliophiles, though, would hardly want their rush to the online sellers to jeopardize the economic security of local bookstores. Yet in our desire to benefit from the personal control that online shopping provides, that is just the sort of unintended effect that might occur.

Of course, not every local business lost to electronic commerce will be sorely missed. In fact, run-of-the-mill shops that we never think twice about would logically be the first to go. Yet we might start to pay attention if those businesses leave behind vacant storefronts and workers without jobs. As taxpayers, we would all bear the burden of this economic disruption.

No doubt, many economists would look at this picture and simply observe that markets are "correcting" themselves by moving sales from "inefficient" offline vendors to their new, streamlined online rivals. But most consumers, I suspect, would see something different here. How many of us would be happy if disintermediation meant higher unemployment rates in our community? Or the closing of favorite neighborhood shops that couldn't compete with well-financed global competitors using the Net to bypass local competition?

The Limits of D.I.Y.

Beyond its effect on local communities and their economic health, the demise of traditional middlemen could also be a mixed blessing for the average consumer. Even from a strictly economic standpoint, the do-it-yourself ethic of electronic commerce could be disadvantageous.

Consider online investing. It may seem like a thrill to trade directly without having to go through a broker. In the near term, online investors should also save money on commissions. But this could actually blind us to the dangers of trying to do everything ourselves. Our new online trading services, to begin with, may not be able to keep up with the huge rise in demand, giving us a false sense of control. One online investor, for example, told the Securities and Exchange Commission (SEC) how he meant to buy a hot Internet stock at $15 to $25 a share, but because of delays and the volatility of the market his order was filled at $90 a share, causing him to spend $150,000 more than he had expected. Complaints of this type have surged.[7]

Some institutional authorities have tried to warn investors about the dangers of oversteer in this context. In January 1999, SEC chairman Arthur Levitt cautioned consumers not to let the ease of online investing get to their heads. Day traders, Levitt said, should be prepared to lose their money. He recounted the stories of online investors using student loan money and retirement funds to try to realize quick gains, and reminded them that "it's just as easy, if not more, to lose money through the click of a button as it is to make it."

Others are even more skeptical. One analyst blames online investing for "the biggest speculative bubble we'll ever see in our lives."[8] Another experienced investor likens day trading to mob rule, adding: "Professionals are trying to tell these people that they're playing with dangerous stuff, but people don't listen when they're in a feeding frenzy."[9]

Even those real day traders who treat it as a career and have intricate strategies—like taking advantage of the spread between bid and ask prices—are, in the eyes of many, doing little more than gambling. A state securities regulator calls day trading a "sophisticated slot machine."[10] And a finance professor adds, "It's almost like being at the blackjack table."[11]

Day traders profit off of price volatility, so even as they make the market unstable they may retain a sense of invulnerability. And as everyday investors hear stories about day traders cleaning up, they may assume that they too can earn a bundle by trading for themselves on the Net. The problem is not only that they might lose a lot of money, but that they could inflate the stock market to such unrealistic levels that it can only come crashing down.

Outside the financial markets, the risks to individuals of bypassing middlemen may seem smaller, but again consumers have to be careful about how much disintermediation is too much. In part that means learning to distinguish between disintermediation and reintermediation. As noted in Part One, buying from Amazon.com or Virtual Vineyards is not just about the decline of middlemen; it's also about the rise of new ones.

These new digital intermediaries can be especially helpful in reducing search and transaction costs—the intangible costs associated with finding and purchasing a product. Yet the irony is that many prophets of electronic commerce seem to be urging consumers to bypass middlemen altogether in order to achieve "friction-free capitalism."[12] If we followed this advice, we would be left to fend for ourselves in an increasingly complex marketplace. Without the help of an intermediary, it would take us much longer to find the right book or airplane ticket or car. It would be costly and difficult to compare prices and even to make purchases with confidence. Search and transaction costs would increase rather than decrease. Even if absolute prices were lower than those found offline, higher search and transaction costs would offset that advantage. At a certain point, disintermediation would be more of a hassle than it was worth.

If we stop to think about it, none of this should be surprising. We don't need fancy economic theories to appreciate the ways in which we rely on commercial middlemen, trusting them to help us decide what we want and to offer us quality merchandise and service. And these are only the most obvious benefits that we derive from a system of commerce that includes middlemen. Middlemen also play many important societal roles that we may not have recognized before, particularly when it comes to government regulation in the public interest. Indeed, it is our digitally enabled ability to do away with these intermediaries that may show us just how valuable and necessary some of them are.

Agents of the Public Good

It may not be a popular example, but the collection of sales tax may provide the clearest illustration of how essential middlemen are to the public good. A main purpose of government, of course, is to raise revenue for public endeavors such as education, social services, safety, and infrastructure. Much of that money comes from sales taxes, which state governments collect from the final seller of a product. In the case of music, for example, taxing authorities don't lean on the record company or the music consumer, but on the record store. They rely on the middleman to collect a share of each transaction and remit it to the state. Why? For the same reason we rely on intermediaries to achieve many collective goals. It's efficient, since intermediaries are generally identifiable and accountable. As central players in commercial transactions, they are ideally suited to act as agents for the common good.

> Mesmerized by dreams of friction-free capitalism, we may miss the important roles that "friction" plays.

Another area where we traditionally rely on intermediaries is in protecting the integrity of transactions and preventing fraud. This is true, for example, in the context of banking and financial markets. Securities law requires brokers like Uncle Max to take some responsibility for the legitimacy of sales in which they act as middlemen. Underwriters bear the burden of due diligence to make sure that stock offerings are valid. These intermediaries also are responsible for disclosure of relevant information to consumers.[13] The issuers of securities, of course, retain responsibilities of their own to customers. But, as with taxation, we have recognized as a society that it makes sense to deputize middlemen as guardians of the public interest. All the more reason, then, that we need to be careful about when and how we bypass intermediaries.

The same can be seen in the way we depend on intermediaries to promote important values such as competence, truth, and safety. We require professionals such as doctors and lawyers to be licensed so that consumers do not assume the full burden of determining whether they can expect competent assistance from a provider of an essential service such as health care or legal advice. Similarly, we impose liability on publications that print defamatory statements, as opposed to just the writers of those statements, as a way of getting editors and publishers to root out those false statements that might damage an individual's reputation. And our legal system is structured so that middlemen play a major role in maintaining safety. Sometimes this is obvious, as when gun dealers are required to do background checks for criminal history. But there are more subtle ways that middlemen are used to enhance safety.

Under prevailing tort law doctrine, for example, retailers are liable for harm caused by the products they sell. This is true even if the person injured did not purchase the product directly from the seller. As the great jurist Benjamin Cardozo noted in 1931, the law's placement of this heavy burden on the seller provides social utility.[14] It encourages retailers to make sure that the products they sell are safe.[15] In fact, tort law provides that even intermediate sellers in a chain of commerce may be liable for unsafe products. Thus, if manufacturer A sells a widget to distributor B, and B sells it to wholesaler C, and C sells it to retailer D, and D sells it to consumer E, everyone from A through D is responsible for the safety of that widget.[16] If E is injured by a defect in the widget, he can sue A, B, C, and D. This is a novel way of protecting public safety and one that depends, perhaps in ways we have not realized, on the existence of middlemen.

In each of these instances where we rely on middlemen to achieve public aims, the effect of disintermediation could be devastating. For example, as electronic commerce grows and consumers sidestep middlemen to purchase products directly from manufacturers, it will be difficult for states to collect sales taxes. Amazon, for example, does not collect a sales tax from its customers.[17] And this is perfectly legal. Today, when you buy something over the telephone or the Net and have it shipped to you, you generally don't have to pay sales tax unless the vendor has a presence in your state. (Many states have provisions that require remote purchasers who don't pay sales tax to pay a use tax instead; but these use taxes are rarely paid.[18])

In light of this existing deficiency in the tax system, states and localities are justifiably wondering how they will collect their fair share of revenue as more commerce moves to the Internet. For even if existing sales taxes apply, collection of those taxes on Internet-based transactions would seem to be a vexing task.[19] The National Governors' Association predicts that the loss in revenue will amount to $12 billion by 2001.[20] That's a lot of money for schools and communities to make do without—and a good reason for us to think about the unexpected ways in which commercial middlemen may, in fact, be beneficial actors to have around.

Mesmerized by dreams of friction-free capitalism, what we may miss is the important role that "friction" plays in our society. It is not just economic inefficiency that we will jettison and be rid of. It is businesses and jobs, support for local economies, and regulation that protects safety and funds education and law enforcement.

E ven with a vivid imagination, it is not easy to grasp just how the Net is allowing individuals to take greater control of politics. To many people, voting online in permanent plebiscites may seem far-fetched, while using the Net to track a bill in Congress may be a bit underwhelming. A web site run by an organization called ParoleWatch, though, demonstrates much of what is possible when it comes to in-dividuals using interactive technology to transform politics—and what might go wrong.[1]

The ParoleWatch site is about citizens getting tough on crime—not by voting for law-and-order candidates, but by actually steering the criminal justice system themselves. The mechanics of the site are explained in an opening message from ParoleWatch advisory board member Marc Klaas, father of Polly Klaas, the twelve-year-old girl who was brutally murdered in 1993 by career criminal and parolee Richard Allen Davis:

> In one of the most innovative and practical uses of the Internet yet devised, ParoleWatch advises the public when dangerous felons are coming up for parole. Finally, interactive democracy enters the murky world of our criminal justice system! Finally, citizens can make a difference by submitting their opinions regarding the early release of violent felons! Finally, our interests are represented and our voice is heard! . . .
>
> By being interactive and proactive, ParoleWatch will be a cen-

tral clearinghouse where you can instantly find out about potential parolees in your area. You'll be able to get details about these convicted felons, including information on their crimes, the length of their sentences and parole eligibility dates.

Then it's your turn to take action. You'll be able to send petitions to the Parole Board. Or send email. In other words, ParoleWatch will EMPOWER YOU TO MAKE YOUR VOICE HEARD.

The goal, according to ParoleWatch, is to initiate "a revolution that will transform the nation's criminal justice system."

Upon visiting the site, it became clear to me how easily this transformation could happen. I started by selecting the area I live in: New York City. Then I clicked on a category of criminals: rapists. And instantly, from a database supplied by the New York State Department of Corrections, came a list of hundreds of rapists prosecuted in my neighborhood with the date that they were eligible for parole. Alternatively, I could search by name. Or I could look for all inmates who had committed a particular offense, regardless of where they were prosecuted. Or I could see who was eligible for release in the next few months.

But that was just the beginning. Since ParoleWatch is committed to keeping criminals locked up, the organization makes it easy for the user to contact parole board members, as well as other politicians. Click here to find out who's up for parole; click there to send an email to a dozen public officials telling them you'll never forgive them if they let this guy out.

ParoleWatch does a real public service by giving citizens access to data about violent offenders and their release dates. Previously it would have been cumbersome and expensive to obtain this information. Now interested citizens can stay informed with little hassle or cost. This information is presumably of particular value to victims and their families, who might be comforted to know the whereabouts of those who have done grievous harm to them in the past.

ParoleWatch does something more, though. It prompts people to "take action." And it encourages them to do so based on a very limited view of each case. For each case, ParoleWatch posts information about the crime and statements by the victims and their relatives.[2] If this is all that one knows about a prisoner, who wouldn't vote to keep him behind bars?

For example, when I received the list of rapists in my area, I clicked on the name of one inmate—call him John Smith. Listed there was Smith's offense, his potential date of release, and his prior criminal record. Based on this information alone, ParoleWatch expects me to be able to form a reasoned judgment about whether Smith should be released. Parole decisions, however, are supposed to be based on other factors—notably, Smith's conduct in prison and the likelihood that he can be let out without posing a risk to society.

The nature of Smith's crime, committed years earlier, is relevant, but more so earlier on at the sentencing phase. A repeat offender or one who commits a particularly heinous crime, for example, may justifiably receive a sentence with no possibility of parole. But an offender like Smith who convinces a judge or jury that he should not receive such a harsh sentence is entitled to a parole decision that is not simply a public referendum on whether his original sentence was too lenient. ParoleWatch, in other words, pushes individuals to second-guess the judgment of the criminal justice system—and to do so based on reading a few sentences and clicking "Keep Smith in jail."

ParoleWatch's tough stance on violent felons is certainly understandable. It is a victims' rights organization and is, of course, fully entitled to express its views on crime and safety. And again, there is much about its web site that deserves praise. Law-and-order advocates, moreover, have always had the ability to lobby parole boards. The difference, though, is in ParoleWatch's sophisticated harnessing of new technology. It uses the pinpoint accuracy of a computer database to match citizens in certain neighborhoods with felons who might be released there, and it relies on the Net's interactivity to give folks a free and easy way to voice their opposition—without having to take time to become well informed, let alone to hear the views of others.

If citizens were urged to express their views after being exposed to more thorough and evenhanded information, this level of interest in our criminal justice system would be welcome. But without such a complete view, justice will not be advanced by this type of resource. In fact, our parole system could be undermined by such an effort to get uninformed individuals to pressure public officials one way or another. The ParoleWatch web site, in other words, is an unlikely yet revealing example of the danger of oversteer in the political arena—of individuals trying to take too much control from government officials.

Mass Government

No one wants to argue that there is a downside to giving individuals more influence over political affairs and policymaking. On the contrary, our democracy is properly rooted in the idea that citizen participation in politics is the wellspring of a civil society. Without an engaged and active citizenry, democratic institutions become weak and illegitimate. The low voter turnout that has characterized American politics in recent decades, for example, suggests that our republic is a democracy in name only. Presidents are often elected and congressional majorities are routinely determined by the votes of a quarter or less of all eligible voters.[3] Democracy would, therefore, be well served by individuals using the Net to increase their participation in political life.

This assumption, however, must take into account the way in which the control revolution is changing the character of politics. Half a century ago, in assessing "the true boundaries of the people's power," the journalist Walter Lippmann pithily summarized the conundrum of representative democracy. "A mass cannot govern," he said.[4] Well, now we seem poised to test that maxim.

The crux of direct electronic democracy is that individuals can exercise a whole new kind of civic power. It is more than just the ability to cast a vote online or to express our views more easily to elected representatives and career public servants. Rather, it is the opportunity to take more control of the decisions that have been made for us by these public officials. Traditionally, we have relied on them to act as our agents, using their training, experience, time, and judgment to determine social needs and to allocate resources. What choice did we have? There was no way for us to know instantly what was going on in the realm of politics, let alone to assert our preferences directly. Decisions were therefore made for us by others.

Now, though, technology may allow us to make many of these choices for ourselves. We could become not just citizens, but *citizen-governors*—each of us playing a role in governing the distribution of resources, the wielding of state power, and the protection of rights.

Where would we start? Maybe by scrapping parole boards altogether and instead voting directly ourselves on whether a prisoner stays behind bars. Those parole board members, after all, are just intermediaries. And the same goes for many others in the criminal justice system. Why should a few jurors decide the guilt or innocence of a high-profile defendant like O. J. Simpson or Oklahoma City bomber Timothy McVeigh when we could all be electronic jurors? They don't even have any expertise. And we might decide that we know enough from a few unadorned facts to determine what is fair and just.

The same reasoning might occur in any area of legal and political life. Why should we trust a judge when evidence and legal rules can be reviewed by anyone online? For that matter, why should we trust the police? (With the emergence of affordable private surveillance tools and "info war" weapons, we should expect to see an increase in high-tech vigilantism in future years. Already there are stories of vindictive individuals using the Net to get revenge.[5])

> "A mass cannot govern," Walter Lippman said. But now we seem poised to test that maxim.

And then there is the ultimate political question posed by the control revolution: If we can use technology to express our governmental preferences directly, why do we need legislators and bureaucrats at all? So many of them, it seems, are captured by monied interests or are merely feeding their own egos. Why not just bypass them and make the rules ourselves?

Tipping the Balance

There are, as noted in Part One, utopian futurists who actively support this radical disintermediation of politics. Representative democracy, they say, is anachronistic and unnecessary in an age when the consent of the governed can be measured not just indirectly every two or four years, but every minute via interactive media. They believe firmly that, as in Athens, each of us can be a citizen-governor.

And do these extremists have even a remote chance of abolishing Congress and relegating politicians to the ash heap? Probably not.

This, however, should not make us complacent about a more incremental yet also perilous tipping of the balance between representative and direct democracy. Interactive technology is already giving individuals a degree of control over their representatives that is unprecedented in our nation—at least in modern times.

In colonial America, mistrust of authority was so great that the first political representatives were legally bound by voting instructions given to them by their constituents (or at least those few property-owning white males who were eligible to vote). Representation literally meant that politicians would gather in assembly to *re-present* the views of these constituents. A provision guaranteeing this arrangement was proposed for the federal Bill of Rights, but the Founders squarely rejected it.[6] Instead they advocated and built a form of democracy in which representatives would walk a fine line between following the mandate of the people and exercising independent judgment. And despite some excesses in each direction, equal commitment to these two principles has continued to be the touchstone of American democracy.

In recent years, though, this balance has been challenged. Even without a formal system of direct electronic democracy in place, we are already experiencing some of its character. The rise of media handlers, instant polling, and focus groups suggests that politicians are ever more beholden to the public—not to citizen mandates, since that implies a carefully crafted and considered message from the public, but to evanescent poll numbers and to impetuous bombardment by today's tools of political outrage: phone, fax, and email. On many hot-button issues, the prevailing wisdom is that politicians who don't follow along will be out of a job after the next election. Already, then, politicians are being pushed to making decisions that reflect not their own reasoned judgment after careful study of an issue, but the snapshot views of their constituents.

The advocates of direct democracy, therefore, may be succeeding in gradually chipping away at representative democracy. Legislatures full of lawmakers still exist, of course. But too often those leaders seem reduced to being glorified poll tabulators (with, it should be noted, a particular interest in the views of their campaign donors). This rise of what we might call push-button politics—a kind of quasi-direct democracy where citizens manipulate their representatives like puppets on a string—is a worrisome development for all the same reasons that direct democracy itself is dangerous.

Deliberating Matters

First and foremost, our political system presumes that decision-makers will make judgments based on reasoned debate and deliberation. Direct democracy supporters say it is wrong to assume that citizens will not have, or do not now have, the time and energy to deliberate over important political matters. But even a minimal reality check suggests otherwise. Amid the pressures of work and family, how could we possibly hope to master, over the course of a few evenings or a weekend, the history and details of a thousand-page budget bill? (Why, frankly, would we want to?) Keeping up with representative democracy is hard enough.[7] And certainly, most pollsters don't ask respondents to deliberate an issue at length before selecting a yes or no answer.[8]

Second is the issue of expertise. Policy and law should ideally be shaped by trained professionals who have the resources at their disposal to be effective leaders and problem-solvers. Sometimes this is framed as a question of whether citizens are sufficiently informed or thoughtful to make decisions for themselves. But there is no need to condescend. The division of labor in a complex society simply requires that we have lawmakers and civil servants who are appropriately schooled in the arcana of politics, including the ability to negotiate, coalesce, and compromise.

Third, decisionmakers must be independent enough to protect everyone's interests and rights. This was a primary reason that the Founders preferred a representative republic to a direct democracy.[9]

As Madison argued, factionalism—or what we might today call "special interests"—might corrupt the union were it not for the independence of representatives. Citizens would likely be inclined to indulge their own narrow interests, as indeed they sometimes seem to be today when participating in focus groups and polls.

Fourth, we should be concerned about civil liberties and civil rights. Direct democracy advocates often make the mistake of assuming that democracy and majority rule are synonymous. Yet our system of constitutional democracy carefully insulates some rights from the whims of the majority. Without such safeguards, we would live in a less free society. Polls routinely show that, if given the chance, Americans would repeal many of the Constitution's key protections of civil liberties.[10] The Founders understood that majority rule could occasionally be dangerous.[11]

Finally, push-button politics could be particularly hazardous when combined with excessive personalization of experience and disintermediation of the news industry. We have already seen how the control revolution might unexpectedly cause individuals to be guided by narrow self-interest and misled by unreliable information. It would be particularly problematic if we were to try to steer politics from such a warped perspective.

The final form of oversteer I will look at is a case study of what happens when political disintermediation is pushed too far. When society decides that thorny problems such as protecting privacy can be solved by individuals with little or no help from government, we can see vividly how the rhetoric of individual control can actually deprive us of the same.

Chapter 15
Privacy
for Sale

I've got Ted Turner's Social Security number here, along with Rush Limbaugh's home address and a couple of phone numbers for Bob Dole in Kansas. I found this information for free on the Internet in about ten minutes. With a little money and some wily sleuthing, I could probably use this data to get their credit histories, financial records, and maybe even some confidential medical facts.

Anyone online, in fact, could do the same, thanks to the emergence of a sophisticated private surveillance industry that is rapidly overshadowing threats from the state. It was once too expensive for anyone but the government to collect, store, and coordinate data, creating profiles on hundreds of millions of citizens. But the growth of networked computing has allowed data compilers, direct marketers, and list-sellers to gather and sell personal information about practically everyone.[1] The result is a broad and lucrative market for personal information that allows anyone with a buck to find out a whole lot about anyone else, just by trolling around the Internet. It's Orwell meets Adam Smith.

Whatever power individuals are gaining here, though, is nothing compared to the increasing vulnerability we should all feel to electronic snoops, both amateurs and professionals. As President Clinton has noted, "Marketers can follow every aspect of our lives, from the first phone call we make in the morning to the time our security system says we have left the house, to the video camera at the toll booth and the charge slip we have for lunch."[2]

Despite this threat, the prevailing wisdom in the U.S. (including in the Clinton administration) has been that technology will empower individuals to protect their own privacy—with little or no help from government. It's a mistaken presumption, one that persuasively shows the dangers of oversteer in politics: namely, the idea that we can bypass government and take care of everything ourselves.

The Market for Privacy

Traditionally, privacy advocates have responded to data collection with calls for broad federal legislation to replace the scant patchwork of state and federal law that now leaves personal information unprotected. Proposals usually require conspicuous notice of what information is being collected and for what purpose; meaningful and informed consent by consumers (for example, allowing them to "opt in" to data collection rather than having to "opt out"); the ability to access files about oneself and to correct inaccuracies; a scheme of redress for violations; and creation of an independent federal privacy protection agency to enforce compliance.

But with the control revolution's emphasis on individual choice and market-based solutions to social problems, many privacy advocates have taken a different approach. The same way that advocates of direct democracy have called for government representatives to give individuals more control, these advocates have demanded that consumers, not government, be empowered to control the flow of personal information.

To achieve this, they have called for the creation of a "market for privacy"—a market that would compete with or complement the growing market for personal information. A report released in April 1997 by a presidential advisory panel, for example, mentioned "the intriguing possibility that privacy could emerge as a market commodity in the Information Age."[3] Just as there is demand for consumer data among corporations, so there is a counterdemand on the part of individuals to keep that information private. The answer, these privacy advocates claim, is to have consumers embrace this market and bargain with vendors over acceptable rules for data collection and use.

For example, if I'm a real stickler for privacy, I may want to pay more to use an Internet service provider or a web-based service that will guarantee me Level 5 privacy (on a hypothetical 1 to 5 scale where 5 represents a commitment to gather a minimal amount of data). Someone else who doesn't care at all about privacy can pay less to use a Level 1 provider, the kind that sucks up data like a Dustbuster. From the company's standpoint, this makes sense because there is tangible value in that data. If they get it, they charge you less (or give you more); if they don't, they charge more (or provide less).

This, in some sense, is how the Net works today. Web sites generally offer their material for free; in return, users give them personal information. This may mean typing your name, email address, or more in some blank registration field. But web sites can also surreptitiously collect data such as what Internet service provider you use, what site you most recently visited, and what computer and browser you're using. Since the Net is already so geared toward information exchange, some privacy advocates figure they might as well formalize that process in an open market: a market that might extend beyond the Net to every exchange of data—with the stores you shop at, your bank, your doctors, and so on.

This market approach has received support not just from industry groups looking into it but from some leading digital civil liberties organizations. They have supported protocols like P3P (the Platform for Privacy Preferences Project), a technical standard that allows users to negotiate privacy practices with data collectors in a way similar to my Level 1 to 5 example, and TRUSTe, a coalition that rewards privacy-friendly web sites with a sort of Good Housekeeping seal of approval.

> The privatization of privacy protection will create as many dilemmas as it solves, if not more.

The goal of these endeavors is to give individuals more control over personal information, rather than having control be entrusted to government regulation or even industry self-regulation. (Indeed, some advocates believe that the amorphous term "privacy" should be scrapped in favor of "data control.") Vice President Al Gore explained his support for P3P by noting that "it will empower individuals to maintain control over their personal information . . . ."⁴

Such a goal is certainly consistent with the control revolution. Recognizing the value of information as an asset, it seeks to give consumers a kind of property right in their information. Your data and sanctuary are your own; you sell them only if you choose—and you can, at least in theory, choose exactly who knows what about you. This would seem to be better than today's free-for-all, where the few rules that exist are vague and, even worse, data are routinely taken from us by invisible thieves.

Yet, while there is every reason to applaud the idea of individuals working to safeguard their own privacy, expecting them to do so effectively without any help from government is dangerously naïve. It assumes that individuals can use technology and the market to achieve a task of such complexity that it has, to date, confounded most governments. What it amounts to is the privatization of privacy protection, which will likely create as many dilemmas as it solves, if not more.

Red Flags

To begin with, leaving privacy protection solely to individuals could be highly inefficient. The underlying idea, after all, is that individuals will negotiate unique data-collection agreements with different vendors. Though noble attempts to automate this process are underway, they won't be able to account for every situation. All the time and effort we might have to spend dickering over different privacy arrangements, then, could add up to very high transaction costs. In plain terms, it might mean such a hassle that we would wind up with less privacy than if government had simply enforced uniform data-collection rules.⁵

The market for privacy might also create a false sense of comfort, blinding us to certain unforeseeable consequences of dealing in personal data. For example, though a company may faithfully notify me that it collects personal information for direct marketing, I may be exposed to more than just junk-mail annoyance. Inaccurate or incomplete information in databases is routinely used to determine whether someone should be hired, insured, rented to, or given credit. The readily available nature of data can lead to discrimination, harassment, and even physical danger (as a Los Angeles reporter suggested when, using the name of an infamous child murderer, she bought detailed data about 5,000 children from information broker Metromail). In the arm's length transactions of the market, vendors have "no incentive to have you think about these dangers," says Oscar Gandy Jr. of the University of Pennsylvania's Annenberg School for Communication. "We're not going to be fully informed."

There is a related problem of unequal bargaining power. While most companies are less interested in your data than in having you as a customer, certain powerful firms, such as the three major credit reporting agencies, are interested exclusively in your numbers. The credit agencies are essentially an oligopoly, presenting consumers with little real choice in the market. If you don't like the terms of the deal they offer, there's really nowhere else you can go to establish a reputable credit report that will allow you to obtain, say, a checking account or a mortgage.

A good illustration of how the market for privacy might not work as planned can be found in the case of a person arriving at a hospital after a car accident: Is he supposed to haggle over use of his medical data before he's treated? He's not exactly in a great bargaining position. What would he do if he didn't like the proposed terms—refuse treatment and go elsewhere? What about kids browsing the web who stumble upon, say, the Batman Forever site, which asks them to "help Commissioner Gordon with the Gotham census" by answering questions about what products they buy?[6] As Gandy argues, "The fundamental asymmetry between individuals and bureaucratic organizations all but guarantees the failure of the market for personal information."[7]

The poor would be at a particular disadvantage. As companies are able to charge increasingly higher rates for finer shades of privacy, poorer customers who can't afford these premiums will be left more exposed simply by dint of economic disadvantage. Even if the markups are small, a little added privacy may not seem worth it for those with limited disposable income, especially since they are already likely to be monitored by the state if they receive welfare benefits or live in high-crime neighborhoods. (In fact, only 39 percent of Internet users expressed a willingness to pay a markup of more than half a cent on the dollar to assure their privacy, according to TRUSTe.) Do we really want to perpetuate such a system of first- and second-class privacy rights?

The market for privacy would also leave little room for discretion and balance. Like most legal privileges, privacy is not an absolute. It must occasionally yield to other important individual rights and democratic values such as free speech and public accountability. Privacy, in other words, must be understood in a specific social context. And yet the market for privacy encourages individuals to control information about themselves in a way that is never subject to public scrutiny or to consideration of the rights of others. Whereas a legally enforced right is always subject to balancing with other rights, a technologically based scheme of privacy protection may be too rigid. Particularly in conjunction with total filtering and the weakening of social bonds, it is not difficult to see how the market for privacy might become a refuge for those who can afford to pay to hide their secrets from legitimate public scrutiny.[8]

Finally, looming over all of this is a commodification critique, which warns that privacy becomes debased when treated like common property and subjected to market pressures. "This is like asking people to pay to practice freedom of religion or free speech," says University of Washington professor Philip Bereano. "We do not buy and sell civil liberties. This is commodity fetishism. It is capitalism run amok." So it would seem. Yet Bereano's claim may only be right in the case of well-established privacy rights, like the Fourth Amendment right to be free from unreasonable search and seizure and the due process right to make decisions about intimate matters such as contraception and abortion. These constitutional rights, one hopes, cannot be peddled to the highest bidder. The situation is less clear, however, when it comes to personal information.

In part, that's because privacy is not well defined or protected in our legal system. Privacy is not even mentioned in the Constitution, and our courts and legislatures have made it the somewhat insecure stepchild of legal rights. Video rental records, for example, are protected by federal statute (because during Robert Bork's confirmation hearings for the Supreme Court, a reporter tried to obtain and publish a list of videos the failed nominee had rented). But medical information is not protected by federal law (though bills proposing such protection have recently been introduced in Congress).

Recently it's become increasingly clear that privacy protection is a higher priority in Europe than in the U.S. For example, in October 1998 the European Union implemented a strict directive limiting the collection and distribution of personal data. Indeed, the E.U. and the U.S. have been at odds over privacy protection, with the Europeans urging American government and business to beef up their lax privacy standards.[9]

The Europeans can draw on any number of resources to chasten our leaders. Many instruments of international law recognize that privacy is a fundamental human right. It is also a core value that protects dignity, autonomy, solitude, and the way we present ourselves to the world. As Justice Louis Brandeis wrote in a 1928 dissent, echoing an idea he expressed in a celebrated article he coauthored in 1890, "the right to be let alone" is "the right most valued by civilized men."[10] In this view, privacy attains special status: Just as we don't allow people to sell their vote, their body parts, or themselves into slavery, maybe we shouldn't allow them to sell their privacy.

But does this mean that I shouldn't be able to trade my own data for money or services? The market-failure problems noted above are certainly red flags. Stanford law professor Margaret Jane Radin points out that such concerns have led society to prevent other kinds of bargaining.[11] A landlord, for example, is legally required to keep a rented apartment habitable; he can't ask the renter to waive that requirement in exchange for reduced rent. Similarly, a company can't sell a toaster at a $5 discount to a buyer who agrees not to sue in the event that a product defect causes her to be injured.

What's clear is that the market for privacy won't do away with the need for new statutory protections and government oversight. It certainly won't give consumers the upper hand against the masterminds of data collection. If anything, it will further reduce privacy from an assumed right to the unceremonious status of a commodity. Folks like

Ted Turner, Rush Limbaugh, and Bob Dole will likely pay to keep meddlers from getting access to their confidential information. But what about the rest of us? If privacy is for sale, will we peddle our digits or save our data souls?

In the last seven chapters, I have argued that we must be careful not to exercise our new individual control recklessly. Not only might such excess cause us to run aground on the shoals of self-indulgence, but it would also give institutional actors more reason to resist sharing authority. If we want governments, for example, to accept the individual power that technology makes possible, then we need to demonstrate a capacity for balance and responsibility. A healthy sense of balance will also help us to recognize the dangers of oversteer. Achieving that balance is the subject of the next and final part of the book.

Part Four

Balance

"Imagination reveals itself in the balance or reconciliation of opposite or discordant qualities."

Coleridge, *Biographia Literaria*

The control revolution presents us with great opportunities, yet could also produce unexpected—and unwanted—results. What can be done to minimize the likelihood of negative outcomes? Or, more optimistically, how do we use our new power to expand rather than narrow our horizons, to strengthen rather than diminish democratic values and fundamental rights, to find the right middlemen rather than trying to do it all ourselves?

How, in other words, can we make the revolution turn out right?

At the most fundamental level, a winning strategy must harmonize three goals: First, allow individuals to take advantage of the empowerment, choice, and control that emerging technologies like the Net make possible. Second, identify unwarranted institutional resistance to that new individual power. And third, prevent oversteer.

Specifically, there are six broad areas on which we should focus. Each requires balance between competing values, and each corresponds to one of the chapters that follow:

- rules and contexts
- convenience and choice
- power and delegation
- order and chaos
- individual and community
- markets and government

To reconcile these values and achieve balance, we must, like engineers, get the weights and tensions just right. But we also have to rethink some conventional wisdom about individual power, freedom, governance, and obligation. New technology can bestow great privileges upon each of us. The test is whether we will shoulder the burdens that accompany them. In part, that will mean recognizing the importance of community and collective action as counterweights to both institutional power and individual control.

Chapter 16

Mapping Principles (Rules and Contexts)

Since the Internet's emergence, a common inquiry in public policy circles has been the "metaphor" question. In trying to figure out what rules should apply to computer networks such as the Internet, lawmakers and policy analysts want to know: What's the right metaphor? Is content on the Internet like printed material, which is generally immune from government regulation? Or is the Internet more similar to radio or television, which traditionally have been regulated because channels of communication are scarce and expensive? Or is it most like telephones and the mail, to which the rules of common carriage have been applied, ensuring low-cost, universal service?

This reasoning by analogy is typical of how the law treats any technology at its inception: the automobile initially is governed by the law of the horse and carriage, the telephone is compared to the telegraph, television to radio, cable to over-the-air broadcast, and so on. In each case, the goal is to fit a technology within an existing legal regime; it would seem odd to begin any other way. As a result, lawmakers will dutifully compare the code features of the Internet to those of other media, trying to figure out whether it is most similar to print, broadcast, or common carriage.[1]

Yet the malleability of the Net means that it can, in some ways, resemble each of those formats—or none of them at all. Moreover, the problem with simply comparing the Net to other communications media is that it fails to take into account the new context that this technology is fostering.

While the same might be said of any new communications medium, the Net appears to be changing the existing social and political landscape faster and more substantially than any new technology in recent memory. Indeed, the scope of change has led some observers to conclude that old rules and regulations don't work anymore and should be scrapped. We have seen, for example, that some information owners believe that digital technology makes copyright law useless, while the FBI says advances in encryption require new rules regarding access to communications for law enforcement.

There is a tension, then, between two competing values: (1) for the sake of consistency, apply existing rules, and (2) in light of a new technological context, devise new rules.

Thus far in the Internet's short history, both the "existing rules" approach and the "new context" approach have been tried—each with their fair share of problems. A solution to this quandary lies in finding a balance between those two approaches: a way that we might call the "principles-in-context" approach.

The essence of this idea is simple. In figuring out how a new technology like the Net should be governed, we should not be constrained by the form of existing rules; but neither should we start from scratch in reconciling competing interests. Rather, we should borrow from time-tested arrangements to achieve efficient and just results in a different set of circumstances. This generally means taking the *principles* that underlie existing laws and rules and mapping them to fit a new context. Some modification may be necessary. But generally, as the following examples should demonstrate, this approach will be more effective than either rigidly applying old rules or coming up with entirely new ones.

Kids and Sexual Content

Earlier, I explained how governments responding to the ability of minors to find sexual materials online may overreact and try to radically change the code of the Internet. Courts have properly rejected these attempts, but they have also recognized that the underlying interest, keeping certain sexual materials from kids, is a legitimate public concern. The question, then, is: What principle traditionally underlies laws protecting minors in a nondigital world—and how can that principle be translated to the new context of the Net?

In the United States, the constitutional principle is fairly clear: Materials that a community deems inappropriate for minors may lawfully be kept from children so long as the free speech rights of adults are restricted as little as possible and parents retain the right to override the community's judgment with regard to their own kids.[2]

This results in a variety of rules. Where a parent or other guardian is available to supervise a child, for example, no state restriction on content is needed because of the presumption that parents are in the best position to know which materials their children should not see. Still, government can play a role. Public libraries, for example, routinely help parents to choose suitable materials by publishing lists of suggested books for different age groups. These are sometimes called "white lists" (as opposed to "black lists" of inappropriate books).

When a parent is not available to supervise a child—for example, outside the home in a commercial setting, or when a radio broadcast occurs in the afternoon—a more restrictive rule governs. As demonstrated in the example of Sam Ginsberg, to keep sexual content from kids, society requires commercial retailers to check the age of those who want to obtain adult magazines and other materials. Young-looking adults may therefore have to show identification to receive sexual content. Yet courts have (not surprisingly) decided that these are acceptable burdens in order to protect kids. Parents, if they want, can also obtain adult materials for their children.

How do these rules, and the general principle that underlies them, translate to the Internet context? White lists work just as well online as in the print world, and many libraries and other groups have created them by setting up web sites with links to materials that are appropriate for different age groups. As offline, though, this solution only works where a parent is available to supervise a child's use of the Net. For all the other times that a kid is online, how should the child-protection principle play out?

With the Communications Decency Act, the government took the "existing rules" approach to this question and basically tried to graft to the Internet the vague indecency standards that govern radio and television (while upping the ante with a criminal penalty). The Supreme Court struck down the CDA on First Amendment grounds and expressly rejected the government's strategy, noting that the Net was not like broadcast. The Court added that the CDA would have prevented adults from getting access to speech to which they were entitled and prevented parents from overriding the state's decision about what their kids should see.[3]

An alternative strategy proposed by some civil libertarians illustrates the "new context" approach. It calls for government to recognize the novelty of the Net and therefore to refrain from regulating it, instead allowing parents to install indecency-blocking software on their home computers (such as CyberPatrol, CyberSitter, and NetNanny). If these tools were more precise in excluding only material that was clearly inappropriate for children, this might be a promising solution. But thus far, in-depth testing of these tools has shown that they block a lot of content that is legitimate for kids. They also usually keep users (i.e., parents) in the dark about exactly what it is they are screening out, because of concerns about proprietary methods of filtering.[4] And most importantly, since these software packages are costly and not exactly easy to use, one cannot assume that they would be widely adopted.

The "principles-in-context" approach asks: What about having commercial intermediaries online replicate some of the role that Sam Ginsberg plays offline? Congress, as noted in Part Two, appears to have assumed that there are no such middlemen on the Net. That's why, with the CDA, it forced individuals themselves to develop and use interfaces to screen out material that might be harmful to minors—a burden that the Supreme Court found both technically impossible and constitutionally problematic.[5] But the Congress and the Court were failing to account for the malleability of the Net, which means that intermediaries could probably be found to handle this job. The question is whether, as a matter of law and policy, we would want them to do so.

On the one hand, asking intermediaries to assist in keeping sexual materials from kids could be more efficient and speech-protective than the government's CDA strategy, since individual speakers would not have to bear the cost of screening or worry whether their material was safe for kids or not. On the other hand, it might be cheaper, more precise, and less cumbersome than voluntary use of black-list programs. Who, though, would this gatekeeper be?

The most appropriate intermediaries to deputize might be the manufacturers of browser software—Microsoft and Netscape—since it is their technology that allows a Net user to encounter material stored on a distant web server. In a sense, like Sam Ginsberg, they hand material over to people. Therefore, they might be required to give adults access to the full Internet while steering minors to the equivalent of the kids' section in a store.

How would they do this? After all, Ginsberg can tell pretty easily who's a kid and who's not (or at least who needs to be asked for ID), but Microsoft and Netscape have no idea how old their customers are and therefore no idea who should get access to what. They could find this information out easily, though, using a technology known as a digital certificate—a kind of virtual ID card that can be embedded into browser software to let Internet users identify themselves during online transactions. The browser companies could use digital certificates to establish one simple fact: whether a user was a minor or an adult. (To protect the privacy of the user, no other information would be collected.)

Once they had this age information, Microsoft and Netscape would know who was entitled to use a regular browser, which would give unrestricted access to the Net, and who was only entitled to a new product called a "kid browser." An adult who established her age once online—with a credit card or other form of ID—would download a regular browser to be used from there on. The default browser available on new computers, though, would be a kid browser. Minors, therefore, would have to use a kid browser (unless their parents gave them access to a regular browser, as a parent can do with any magazine in Ginsberg's store).

Who would decide what one could see with a kid browser? In physical space, Sam Ginsberg makes this judgment call, but (fortunately) there would be no way for the browser companies to do so with an endless Internet to evaluate. So other strategies would have to be considered. One would be to have kid browsers give access to a variety of white-list sites online; this would make web access for kids analogous to visiting a children's library. A more permissive strategy would be to have kid browsers give access to all sites except those voluntarily blocked by purveyors of adult materials; but this might be subject to abuse or just reasonable disagreement about what's appropriate for kids. The best option might be to create one kid browser for each strategy and let parents decide which they think is most appropriate for their kids, depending on their age and maturity.[6] (Kid browsers could even be made age appropriate, so that a fifteen-year-old could be given broader access by her parents than a twelve-year-old would get.)

Requiring companies like Microsoft and Netscape to create kid browsers would simply be a way of recognizing that commercial middlemen have a role to play in protecting the public interest. Like Sam Ginsberg, they owe something to the community. Mapping Ginsberg's role to the context of the Net would take some modification, though, in ways that actually turn out to be constitutionally preferable. Unlike Ginsberg, the browser companies wouldn't have to make any judgment about what was appropriate for kids. They also wouldn't be subject to criminal liability if kids did get access to inappropriate materials (the requirement that they create kid browsers would simply be a civil regulation).

With a variety of kid browsers to choose from and the option of overriding the system to let their children use an unrestricted browser, parents would have more control than they generally do over what their kids see. For adult users, having to establish their age once to download an unrestricted browser would be a very minor obligation.[7] (Compared to offline interactions, it would be less burdensome for young-looking adults who are constantly required to show ID to get access to adult materials). A kid browser scheme would be preferable in terms of cost, effort, and efficiency to having parents install their own blacklist software. And it would certainly be constitutionally superior to the CDA, since adults would have unrestricted access to content on the Net.

Expectations of Privacy

The value of applying principles in context can be seen also in the arena of encryption, secrecy, and law enforcement. A fundamental principle of America's constitutional system is that when government officials investigate criminal activity, they must also respect citizen privacy. This principle has traditionally been expressed in specific rules: Law enforcement officials may intercept private communications if they follow certain procedures. To wiretap a phone and listen in on a conversation, police must prepare a sworn statement explaining why they have probable cause to investigate a person, and they must get a magistrate to approve the search. Failure to comply with this process may cause a court to suppress any evidence obtained. In short, government can get access to secrets if it goes by the book. And citizens can, at least in theory, rely on courts to protect them if the cops cut corners.

Does this traditional set of rules map well to the arena of digital communications? Most strong encryption advocates say yes, arguing that law enforcement can still do its job under "existing rules." Law enforcement officials, of course, disagree, noting that even if they get a warrant and intercept a message, it will be undecipherable if scrambled with strong encryption. Here, then, the cops take the "new context" approach, arguing that encryption technology is changing the dynamics of secrecy so much as to require a new regulatory arrangement. The prime feature of that scheme is key escrow: requiring encryption users to make available an extra set of deciphering keys so that police can get access to readable communications.

But is *this* a rule that fairly applies the time-tested principle of compromise between citizen privacy and law enforcement? At first, it may seem so. Yet when we consider the context of the Net, it becomes clear that the government's proposed modification runs afoul of the original principle.

It all has to do with our changing expectations of privacy in remote communications. Law enforcement says key escrow is no different from an analog (i.e., nonencrypted) telephone wiretap, since both require a warrant and both are subject to some theoretical possibility of abuse. In fact, it's probably easier to wiretap a phone than it is to get unauthorized access to an escrowed key. Yet precisely because a wiretap can be done by anyone with some cheap surveillance equipment and a little know-how, most users of traditional telephones don't have a high expectation of privacy. Moreover, they generally don't need to, since there is not much at stake in a typical person's phone conversation.

The Net is different. Users of computer networks are likely to transmit not just everyday chat, but sensitive personal and commercial information such as financial records, medical files, legal documents, and so on. Even if phone users were to speak about such matters, their speech is ephemeral. By contrast, Net users often send *and store* sensitive documents on computer networks. With strong encryption, they can do so with true security. Without it, interlopers—be they rogue government agents or private spies—can get access not just to gossip but to highly personal information. The difference between key escrow and a wiretap, then, is that the expectation of privacy in the former context is substantially higher than in the latter.

To their credit, law enforcement officials are correct about at least one thing: Strong encryption does give individuals more power to keep secrets from government (and others) than they have otherwise had. Yet because the context of the Net means that users are substantially more vulnerable in terms of the information that is at stake, depriving them of strong encryption would actually give them *less* control and security—and less power relative to government—than they have previously had in other contexts.

> The suppleness of code means that it can be altered in positive as well as negative ways.

No viable rule has emerged that would perfectly replicate the previous compromise between law enforcement and citizen privacy. For now, then, a choice must be made: In translating the law enforcement/privacy principle to a new context, do we choose a rule that favors individuals or one that favors government?

Against the backdrop of an increased need for privacy in remote communications, the right thing to do is to err on the side of giving individuals more security. That means government should either scrap key escrow or defer to cryptography experts who would assist in creating an escrow architecture geared more toward safeguarding privacy than those that have been proposed. The suppleness of code, after all, means that it can be altered not just in regressive ways, but in positive ways.

This is not to say that strong encryption doesn't present problems for society. Beyond its use by terrorists and child pornographers, strong encryption could be used more mundanely to hide economic activity that should by law be taxed. (Radical libertarians love the notion of making untraceable electronic deposits to off-shore bank accounts, frustrating the efforts of tax collectors.) The consequences would be felt by everyone: every tax dollar not collected, after all, has to be made up by innocent taxpayers or accounted for with cuts in public services.

Yet the answer to such a potential dilemma, and to others, is not to reflexively deny individuals strong encryption, but to pursue other methods of law enforcement. It is, in fact, particularly in the interest of encryption proponents to work with law enforcement to figure out ways in which our communities can be protected without having institutional powers unnecessarily restrict privacy or the use of emerging technologies. In fact, with its own use of new technology, law enforcement should have many more investigative advantages that will help it to enhance public safety without diminishing privacy rights.

Creative Breathing Room

In the case of copyright and creativity, applying principles in context is again crucial to finding the right rule to reconcile competing interests. The principle at stake is also one of constitutional dimensions: How do we encourage and reward the creation of original works while at the same time allowing the public to benefit from those creations? Traditionally, copyright law has given authors and other creators a limited exclusive monopoly in their work so that they can charge for its use and make a living as writers and artists. This in itself is beneficial to society. Yet the law has also recognized the need, in an open society, to give the public certain limited opportunities to use copyrighted work—for example, for personal use or to critique a work. There is an equilibrium here between rights of property and free speech.

As I explained in Part Two, powerful copyright owners in the entertainment, publishing, and software industries believe that digital technology, including the Internet, is creating a new set of circumstances that could deprive them of their most lucrative assets. Smaller companies and individual writers and artists also are concerned that their source of income may be jeopardized. And these fears are justified, because new technology does give individuals the ability to copy and distribute protected materials with far greater ease than before, usually with impunity. Adhering to the "existing rules," then, is probably inadequate.

Yet the new rules of information protection now being put into place by government and copyright owners go too far. They are expanding copyright law and changing the code of digital technology to require use of information-protection schemes—such as trusted systems and online clickwrap agreements—that could unduly limit the rights of users, while making more and more information available only on a pay-per-view basis. Every information transaction that uses digital technology might be regulated, monitored, and controlled.

The rules for this new context, in other words, are too perfect. In contrast to the balance underlying copyright law, there is no lenience that allows for personal and critical uses of protected works—in other words, no guarantee of fair use.[8] The timeless compromise between protecting works and making them publicly available is therefore in danger of coming undone. These new rules respect one element of the digital context: the ability of the Net to be a giant copying machine. But they ignore another: the sometimes unforgiving and unpredictable precision of technology. Just as filtering technology can unexpectedly cause the benefits of personalization to be squandered, so the new information protection schemes can backfire and jeopardize the important goal of preserving copyright *and* a rich public domain.

The remedy to this situation lies in modifying the new rules of information protection to allow for some of the breathing room that has always been a part of copyright law. Applying this principle in context means that the code of trusted systems and clickwrap contracts must be altered to preserve the ability of individuals to copy and otherwise use a work for a few socially beneficial purposes—parody, commentary, personal use—that would not unduly interfere with society's overall goal of encouraging creativity.

Government should recognize the importance of this breathing room, by adopting a rule analogous to fair use that might be known as "fair hacking" or "fair breach." This would give individuals the right in certain limited situations to circumvent technological protections of information or to ignore the provisions in clickwrap contracts. If legislatures fail to enact such exceptions, courts should find that constitutional principles, including the First Amendment, require that they be recognized.

Chapter 17

Shattering Illusions (Convenience and Choice)

Learning to distinguish the illusion of control from the real thing is another crucial part of achieving balance. Earlier we saw how some media and technology companies may try to convince us we are in charge even as they restrict our choices. In this effort, they will often mask a lack of choice by appealing to individuals' natural desire for convenience and simplicity.

To many people, the Net and computers are still foreign—and this will remain the case for some time, as new users continue to gravitate to the Net. These people may feel more comfortable having access to preselected content and commerce links such as those presented by America Online or Microsoft. Or they may want to use an interface like WebTV's that makes the Net more like television. These tools let users connect more easily to familiar brand-name sources of information and commerce.

Certainly there is much to be said for reducing the confusion of the Net. If users don't feel comfortable online, they won't be able to benefit from the new autonomy that it offers. Making the Net accessible—in terms of price and ease of use—must therefore continue to be a high priority.

There are important trade-offs, however, between short-term convenience and real choice in the long term. In Part Two, for example, I argued that push technologies and appliances like WebTV can tame the Internet, but at the same time diminish important code features such as interactivity. And more generally, when icons for big corporations are the first thing we see on our computer screens, it's unlikely that we'll go searching for alternative sources of news, entertainment, and products.

There is nothing wrong with *choosing* to visit Microsoft's content and commerce partners; in fact, they may offer exactly what you're looking for. But relying solely, or even mostly, on the selections provided by big gatekeepers can cause users to be unaware of their options. Indeed, a fully informed choice can only be made when you have some broader sense of what is available online. It's as if you were asked to choose a magazine from a newsstand that only displayed three different publications. Others might be available behind the counter, but if you didn't know of their existence it's unlikely that you'd have the foresight to ask the newsstand proprietor for one of those titles. And even if you did have some general idea that other publications were available, the public display of only three titles might well persuade you to read one of those instead. (This, of course, is why publishers jockey to have their magazines displayed prominently by distributors.)

Following the selections of major gatekeepers, then, may seem harmless. Yet it can erode the idea of individual control based on fully informed choices—not to mention the fleeting

> Users should feel emboldened not just to take advantage of cutting-edge technologies, but to reject them.

hope that on the Net everyone's voice will be roughly equal, or at least more so than in traditional media. Users may still have access to an endless amount of information online. But in an environment of cutthroat competition for audiences and eyeballs, individuals, nonprofits, and small commercial outlets will have a difficult time capturing an audience online.[1] They probably won't wind up on the Windows channel bar or among the recommended sites on AOL's first screen. Indeed, without the brand recognition or the advertising budget to compete with the big online players, they will likely be about as prominent as the outcasts on public access cable or ham radio. At a minimum, many of the small, alternative commercial sites might not survive, and this could affect the diversity of the Net, leaving us with fewer options to choose from.

Fewer options, though, in comparison to what? Certainly, in terms of information sources, we will still have more to choose from than in the comparatively anemic world of radio and TV. Many average users will therefore conclude that they can afford a little less choice in return for the simplified access provided by interfaces like WebTV or the channel bar on the Microsoft Windows desktop. They may also not see the televisionlike nature of push interfaces as a problem. Indeed, they may conclude that Liebling's revenge—the ability of individuals to produce information, rather than just being passive consumers of it—is a feature that matters more to experienced users than to them.

And perhaps they are right. The decision is theirs to make. Users, in fact, should feel emboldened not just to take advantage of cutting-edge technologies, but to say no to them—whether because of convenience or any other reason. Every user should recognize, though, what's at stake in the selection of an important tool such as a gateway to the Net.

The Antitrust Reality

The tension between convenience and choice is evident also in Microsoft's controversial decision to bundle certain software products. In its antitrust actions against Microsoft, the U.S. government alleged that the company engaged in anticompetitive business practices when it integrated its browser, Internet Explorer, into the Windows operating system. In its defense, Microsoft argued that this integration made it easier for many consumers to use PCs and the Internet. The truth is, both claims were right.

Folding the browser into Windows meant that users could move effortlessly from their own hard drives to any resource online using a single software application. Moreover, Microsoft's bundling of programs in software suites like Office, another allegedly anticompetitive practice,[2] allows for easy shuttling between a word processor, a spreadsheet, and a slide-presentation program.

The problem is that these sources of convenience can also restrict, or even do away with, consumer choice in the long term. With Microsoft's browser, Internet Explorer, integrated into Windows, people will undoubtedly find it easier to use Explorer than Netscape Navigator or any other browser. Competition in the browser market will likely become even more feeble than it has been, possibly leaving us without any viable alternatives to Explorer. The likelihood of this result is evident in the fact that Netscape has lost market share precipitously since Microsoft began bundling Explorer and Windows in 1996. (By the time Netscape was purchased in late 1998 by America Online, its share of the browser market had fallen from more than 80 percent to less than 50 percent.)[3]

Similarly, while it may be convenient to have one software suite that includes a word processor, a spreadsheet, and a slide-presentation program, the combination of these products will make it difficult for any competitor to invest money in developing and marketing, say, a stand-alone word processor or spreadsheet. Whatever efficiency we may gain from Microsoft's bundling of software, then, may be outweighed by the loss of innovation that occurs when rival software companies see no real opportunity to compete.[4]

> With all the Net offers, it will be easy to assume that a small degree of increased control is the equivalent of true self-determination.

Even more important, if Microsoft can dominate the browser market, it can stave off new competition to its Windows monopoly that might come from "platform independent" technologies like network computers or Sun Microsystems's Java programming language. This is significant because these alternatives to Windows could substantially change the way computing and the Internet work (more so than having access to a wide variety of browser programs or word processors).

The antitrust suits against Microsoft, of course, sought to remedy this situation.[5] Regardless of their outcome, though, it is already clear that the federal and state governments have done a considerable public service by initiating these legal actions. The Justice Department, in particular, has shined a spotlight on Microsoft practices that, illegal or not, have raised eyebrows among proponents of strong market competition—including some who are traditionally foes of antitrust law. As a result, industry players, the press, and the public are likely to keep a closer watch on the practices of Microsoft and other market leaders.

Choosing Choice

The antitrust case against Microsoft also appears to have raised awareness among computer users about the importance of making technological decisions from a critical perspective. Evidence of this might be seen in the fact that, during the antitrust suit, support has grown substantially among computer professionals and hobbyists for "open source" software such as the Linux operating system.

Open-source software, also known as freeware, is generally developed voluntarily by programmers and distributed for free. It's called open source because the underlying source code, which usually is proprietary and hidden from users, is made freely available. In fact, individual programmers can add on to and manipulate the source code, and changes can be submitted to loosely knit groups of developers that decide whether the changes should be officially added.

Open source is, in many ways, a perfect software paradigm for the control revolution. Although it is sometimes not as easy to use as proprietary software, it gives individuals more control over the computing experience and promotes cooperation simply for the purpose of building a better and more useful product. (Microsoft, not surprisingly, sees it as a threat.[6])

Open source has always been an important component of the software landscape—particularly in terms of Internet protocols and web site design—and its growth is a healthy sign for the computing industry. But it may not be for everyone. Some vendors are beginning to sell convenience packages, including manuals and customer service, to accompany this free software, and projects are also underway to make open source more user-friendly. In the end, though, the trajectory of open-source software will probably only confirm how difficult it is to preserve individual control and choice in an atmosphere where convenience is most prized. If open source continues to flourish, it

will be a testament to the efforts of a relatively small group of supporters working against the odds.

The federal government, though, could help tip those odds. It could:

- *Encourage innovation and competition by supporting the development
  of open-source software:* Recalling the federal government's role in
  supporting the research that produced the Internet, Congress should
  authorize the National Science Foundation to support the development of nonproprietary software in order to encourage innovation and competition in the software industry. This might be a
  particularly appropriate course if the government loses its antitrust
  suit against Microsoft or if it wins and the judicial remedy still fails
  to open up the software market.

- *Use its procurement power to create a more competitive software market:* The federal government could, for example, install Linux or
  other open-source software on some share (perhaps 10 or 15 percent) of all government computers, at least in a trial run. State and
  local governments could do the same. This would substantially bolster acceptance of open-source software, giving it an opportunity
  to take hold in the corporate and personal markets. The federal government has used its procurement power in the past for a variety of
  purposes (from encouraging racial and gender diversity in certain
  industries to promoting technologies such as its preferred encryption standard). This time, it would be doing so for a cause that most
  would support: promoting innovation and competition.

These suggestions, of course, do not preclude other actions. Some
people will commit to seeking out less-traveled paths online or to trying alternatives to Microsoft such as the Apple Macintosh or Linux
operating system. Still others will protest loudly, and urge others do
the same, when large companies deprive them of decisionmaking control.[7]

Even if few of us actually make decisions about technology based
on considerations other than convenience, we should all be aware of
the elusiveness of real freedom of choice. Though we may feel as if we

have control over our technology, the full substance of it may not be there. We may, for example, have what psychologists call "outcome control" but not "agenda control."[8] That is, we may possess control over specific short-term outcomes—should I click x or y?—but no control over the broader agenda of what the Net is about: whether it is diverse or monochromatic, dominated by a few companies or open to robust competition, trivial or enlightening.

With all that the Net offers in the way of personal enrichment, it may be all too easy to assume that a small degree of increased control is the equivalent of true self-determination. The challenge is to ensure that we do not become, as Erich Fromm put it nearly six decades ago, "automatons who live under the illusion of being self-willing individuals."[9]

Chapter 18

In Defense of Middlemen (Power and Delegation)

One way to understand the impact of the control revolution is to appreciate the way it can make each of us the CEO of our own company—call it My Life. We're in charge, free to make choices about every little detail. But as any experienced executive will tell you, the boss who refuses to delegate some decisions to others is not only unhappy but unproductive. Political representatives, news professionals, and commercial middlemen can make awful decisions on our behalf—but we're generally better off with them than without them. A vital component of balance, then, is learning when to choose for ourselves and when to delegate decisionmaking power to others.

Just as a choice between x and y does not necessarily give us genuine control, being in control does not always require choosing for oneself. Indeed, the need for balance between convenience and choice mentioned in the last chapter can be solved in part by delegation: by asking others whom we trust to make decisions on our behalf.

Delegation goes hand in hand with the control revolution because it allows individuals to establish their authority even as they recognize that others may be better suited to carry out certain tasks. Indeed, the more that people are empowered by technology, the more effectively they should be able to parcel out duties, since they will have a broader range of candidates to choose from. We should not, however, delegate and then micromanage our appointees. Rather, we should select them with great care and then regularly review their performance, while otherwise allowing them to exercise their own judgment.

Technology writer Kevin Kelly agrees about the need for some delegation, but thinks we need to give up power to our computers. "Giving machines freedom is the only way we can have intelligent control," Kelly argues.[1] But letting computers make decisions for us would deprive us of the additional wisdom that comes from others whose intelligence—*human* intelligence, that is—we respect.

> Delegating power to trusted intermediaries will make us more free.

What we need are trusted intermediaries: people to whom we entrust certain tasks because we recognize the value of their perspective, their expertise, their time, and their independence.

Delegating power to trusted intermediaries does require some relinquishing of decisionmaking control. But if we select trustees carefully and make sure they continue to live up to our expectations, we don't need to give up anything in the way of personal autonomy. We should, in fact, be more free than if we try to do everything ourselves, and more connected to one another as well. Delegation, after all, is a fundamental building block of community.

Two areas in which we can particularly see the benefits of delegation are news and politics.

News and Truth Watching

It is fashionable these days for people to ridicule the professional news media. Among the digerati, though, one can hear not just ridicule but gleeful predictions of the news industry's demise. "Journalists are standing on the deck of a sinking ship," one Internet entrepreneur told *Wired* magazine. "They can either get into the lifeboats now or go down."[2]

Disintermediation of news and information has its benefits. We can bypass journalists to present our own account of events or challenge them when we think they've gotten the story wrong. Yet we also have to face an information terrain that has few familiar landmarks and many potholes and dead-end streets. The Drudge factor complicates our media environment, requiring each of us to be more discerning and wary in our selection and interpretation of what we read and watch. Certainly it is time for us to abandon the idea, if we haven't done so already, that a fact is true simply because it has been "reported" somewhere. Instead we must dissect the news in much the same way that we interpret a film like *Rashomon*, in which director Akira Kurosawa intentionally presents multiple, inconsistent perspectives on the same event.

This requires that we rely more, not less, on certain trusted intermediaries: not superpersonalized news services, but outlets that put a premium on being right instead of on being first. We need to demand higher standards from our media middlemen, as opposed to the follow-the-pack performance that we have seen of late. We should, in other words, leverage our new power as news consumers to get better results from the media. We should reward them with our attention or our subscription dollars only if they take seriously their job of presenting reliable information in a meaningful context.

This does not mean that we should exalt the traditional media or even give them the benefit of the doubt. Though the Net may seem to supply more dubious information than print and broadcast media, even our journalistic standard-bearers can lead us astray.[3] And the Net, with its speed and interactivity, can be an antidote to that, as we saw in the *Time* cyberporn case. The real problem has less to do with the new media versus the old—an already blurry distinction—and more with what happens to us, the audience, when we're increasingly trying to evaluate a deluge of data for ourselves.

That is why we need top-notch journalists and editors—particularly editors—now more than ever.[4] Only we need them to adapt to the control revolution and the dangers it poses to the integrity of information. That means being more careful about getting the facts right. And it means that we need a new breed of media professionals who will specialize in assessing the accuracy and relevance of information that comes from a variety of media.

When a controversial story arises, these "truth watchers," as we might call them, should provide us with the story behind the story: who were the main sources, how close were they to the action, what bias might they hold? They should periodically review the performance of different media outlets, grading their precision and giving consumers information about their methods and reputation. A decent way to think about it is that we need an information immune system.[5] Truth watchers will help us root out the bad data cells and protect the good. They will be our first line of defense against inaccurate information.

To a degree, these truth watchers already exist. There are organizations like Fairness and Accuracy in Reporting, and publications such as *Columbia Journalism Review* and *Brill's Content*, that specialize in holding the media accountable. But we need more entities like these that will track the accuracy and biases of various information sources. We also need more offerings like Today's Papers, an invaluable daily email service from the online publication *Slate* that summarizes and compares the lead stories of the nation's largest daily newspapers, often with heightened attention to the way that those stories present different accounts of the same events.

One of the great assets of a service like Today's Papers is that it helps to deconstruct the pretense of objectivity in journalism. This may seem at cross purposes with the goal of ensuring accuracy, but the two values are not the same. Accuracy is simply a matter of getting the facts right, to the degree that is humanly possible. Objectivity, on the other hand, suggests there has been no personal or institutional spin on the presentation of a story, which is impossible since even the decision to select one fact over another reflects some subjective choice. Certainly we want a cadre of journalists whose first duty is to report what happened with a minimum of editorializing. But the cloak of objectivity worn by the mainstream media is too often a cover for an unstated agenda. Mild as that agenda might be, it is better for readers to be aware of a news intermediary's perspective. That way, in a competitive market, we can make fully informed choices about the information we receive.[6]

Making such choices involves relying on the reputation of a news outlet. Whether a story appears on National Public Radio or in the *National Enquirer* helps us to determine the amount of credence we give it. Truth watchers, then, must help us to make judgments about the quality and reputation of different information sources.

To earn their stripes as trusted intermediaries, journalists in the digital age also must be dexterous guides, steering us through dense forests of data to show us the few trees we really should see. "The historic function of the news includes understanding what your audience cares about and thinks—but also telling them things they may not know are important to them," says editor and media analyst James Fallows.[7] No matter how smart we are, we can't know in advance all the information that we want to know, let alone need to know.[8]

This is one of the problems with various proposals to do away with information intermediaries. Law professor Bernard J. Hibbitts, for example, has urged legal scholars to stop publishing in student-edited law reviews and instead to post their articles directly on the web.[9] There is a long (and perhaps justified) tradition of dissatisfaction with the practices of student editors, and Hibbitts is right when he says that self-publication would allow scholars to take back control from those students. But the problem is that readers and authors alike would be deprived of the imprimatur provided by the law review. Well-known authors might not need such endorsement, but other writers do—and readers certainly benefit from it.

One solution might be to replace such journals with a different reputational intermediary, a consortium of scholars who would agree to review one another's articles and endorse them as worthy of publication. When a piece received a "ready for publication" mark from three readers, for example, it could be posted by the author on the consortium's web site, perhaps with the names of the three endorsers. This would not be disintermediation so much as a more efficient form of intermediation.[10]

The lesson here that applies to intermediaries generally is that we usually don't need to bypass middlemen so much as we need to reform them—or find new ones. The wisdom of this point should be clear not just in the world of news and scholarship but also in other pursuits, including commerce. With low barriers to entry, markets for products and services will expand and, without some guidance from intermediaries, become increasingly cluttered and difficult to navigate. Appealing as the unmediated interactions of "friction-free capitalism" may be in theory, too much circumvention of middlemen will, as we have seen, only confuse our online interactions and lead to dashed expectations.

In the same way that we will need trusted intermediaries to filter reliable reporting from gossip, then, we will need trusted intermediaries to help us find high-quality commercial goods and personal services online. Again, they should be intermediaries whose judgment and reputation we can count on. Some of these trustees may be traditional commercial outlets who have found a presence online. Others may be new digital intermediaries that aggregate and compare information. In Chapter 21, I will discuss further the need for balance in the commercial arena. It is just worth noting here that sometimes a degree of freedom *from* choice can be as liberating as choice itself.

Politics and Re-Delegating the People's Power

Politicians can, in one sense, rest easy. Though the control revolution would seem to make direct democracy possible—technologically, at least—political middlemen are not going to be put out of business anytime soon. Athens-style unmediated democracy (what Madison called government by citizens "in person") is simply inimical to our political tradition, not to mention the dynamics of everyday life in a complex nation-state.[11]

Yet push-button politics—that form of quasi-direct democracy where representatives are beholden to instant electronic polls—is an increasing reality, and one that poses threats to our constitutional system and our personal well-being. And here politicians should worry, because the solution to this problem lies partly in finding more reliable intermediaries: in other words, better representatives.

As with our potential to bypass journalists, the possibility of circumventing elected leaders should cause us to think long and hard about what we want our political system to look like. The individual control that the Net allows may lure us into believing that direct democracy is a viable option, but we need instead to harness our power toward the opposite goal: strengthening representative democracy.

To start, we need a new understanding of democratic delegation. If our politicians are really going to be trusted intermediaries—peo-

ple whom we authorize to make independent decisions on our behalf—then they must prove to us why they deserve our confidence.
Some officeholders, to be sure, might assume that they need to pander further to polls and focus groups. But this is the exact opposite of
what we should demand of them. Just as a journalist adds no value by
dumping unconfirmed rumors on readers, a politician adds no value
to the quality of his representation if he simply mirrors the immediate desires of the people. We can, after all, do that ourselves with push-
button voting. Politicians must therefore justify themselves as being
something more than just adroit poll-readers. They must represent the
people but be independent of their whims. They must be pillars of
judgment rather than bellwethers of faddish opinion.

To help politicians become effective intermediaries on our behalf,
one thing we must do is relieve them of the absurd burden of constantly raising money to run for office. Not only does this fundraising
eat up precious time that representatives should spend serving us, but
it compromises their ability to serve our best interests, as the money
chase requires them to pander to those who can give the most. We
wouldn't accept such a compromised performance from other trusted
intermediaries (in news, education, or commerce); rather, we would
give them the resources to do the job right. And that is what we should
do with our political middlemen: provide them with full public fi-
nancing to let them focus only on serving us, while enacting strict limits on campaign fundraising and spending to make sure that they are
worthy of our trust.[12]

Encouraging independence on the part of politicians does not mean
that they should be unresponsive to citizens. Constituents and community leaders should still, of course, meet face to face with politicians. And technology could actually help to make remote
communication more effective. Citizens and their representatives
should work together to:

- *Create forms of remote citizen-representative communication that are*
  more robust and thoughtful than instant polls. We should experiment
  with deliberative forums online that would allow us to become
  deeply engaged in issues in an ongoing fashion. Instead of just an-

swering questions—should this budget bill be passed? should that treaty be signed?—citizens would choose an issue or set of issues that they wanted to focus on in some depth. The key is commitment: We would be required to read materials, listen to others, and participate in conversations before expressing our preferences. (This would be consistent with other situations in which we are required to listen before exercising our political rights—for example, a juror cannot vote on a case unless she hears testimony and argument.)

This format of decisionmaking—called deliberative polling by political scientist James Fishkin—has been tried with some success in off-line forums and there are plans to experiment with it online as well.[13] The results of online deliberative polls should not be binding on politicians. In fact, they might just be a good way to increase mutual understanding between citizens and their representatives. However, to show participants that their commitment to democratic deliberation is being taken seriously, the results could also be agenda-setting: that is, a representative group of citizens in a deliberative forum might ask a legislature to address a certain question or issue. Whatever the outcome, if we are going to use the Net as a tool of deliberation, we must develop more sophisticated software tools for ongoing citizen interaction. Anyone who has engaged in an online dialog knows how easily debate can become unstructured, truncated, and ultimately superficial.[14]

There are other ways we can use technology to enhance representative democracy. We can:

- *Inform ourselves so that we are more effective participants in the political process.* Rousseau and Jefferson, often cited as true believers in direct democracy, were actually greater proponents of lifelong civic education than of unmediated self-rule.[15] (Perhaps they recognized that educated citizens would appreciate the benefits of representative democracy.)

- *Establish voter registration online* in order to increase the likelihood that people will vote. Benign as this may seem, many states have gone to great lengths and spent much money to oppose federal efforts to make it easier for citizens to register to vote.[16] Ideally, voter registration should be universal and automatic; at a minimum it should be as easy as visiting a web site.

- *Explore the feasibility of letting individuals vote from home or work on election day* using a secure, authenticated Net connection. Allowing individuals to vote remotely could increase participation among the elderly and the infirm and those who believe they are simply too busy to get to the polls. Using encryption technology, it should be possible to achieve remote voting in a way that satisfies concerns about voter fraud online. Eventually, cheap computer terminals could even replace the expense of voting machines at polling places.[17]

The goal in each of these instances is to increase the number of people who participate in the political system, and the effectiveness of their participation, while maintaining a distinction between direct and representative democracy.

What we don't need is knee-jerk opposition to electronic activism, which is often driven by ignorance. A case in point was a 1997 newspaper column by Cokie and Steve Roberts expressing alarm that the Federal Trade Commission (FTC) was allowing citizens to send "comments" by email on a pending merger.[18] When the FTC and other regulatory agencies take public comments on a pending case or issue, though, they are not just asking for a flood of yes or no replies. Rather, as required by law, they are requesting detailed responses from interested parties—community groups, industry alliances, affected citizens—regarding the merits of a proposed decision. We should therefore applaud the fact that individuals and groups would take the time to learn about such complex issues and that the federal government would welcome use of the Net to expedite the process of review.

The conflict between direct and representative democracy will likely become more contentious as technology continues to empower individuals. Though many of us may take for granted the need for a representative form of democracy, some citizens will remain fundamentally dissatisfied with the idea that, although "we the people" are sovereign, a good part of sovereign decisionmaking power should be delegated to others. A persuasive case will have to be made to citizens that trusting political intermediaries is a fundamental aspect of a balanced approach to the new individual control.

Chapter 19

In Defense of Accidents (Order and Chaos)

In a brief essay entitled "The End of Serendipity," journalist Ted Gup recalls his experience as a boy using the twenty-two-volume *World Book Encyclopedia*. A homework assignment might begin with the goal of looking up *salamanders*, but soon he would be joyfully waylaid by witch hunts in *Salem* and by ancient *Salamis*, where Greeks and Persians battled. Gup then explains how his young sons now use the same encyclopedia, only it's an electronic version that requires just the typing of a word to find an entry: "In a flash, the subject appears on the screen. The search is perfected in a single keystroke—no flipping of pages, no risk of distraction, no unintended consequences. And therein lies the loss."

Despite the efficiency and convenience of such a tool, Gup says, its creators "are inadvertently smothering the opportunity to find what may well be the more important answers—the ones to questions that have not yet even occurred to us." Ultimately, he fantasizes about a computer virus that would cause his sons to type "salamander" and find not just that topic but a random array of others, giving them "the sheer joy of finding what they have not sought."[1]

Gup's composition is a delightful ode to accidents—the small, unplanned occurrences that take us by surprise and uncover curious gems we would not otherwise have looked for or found. Sometimes these gems are people, sometimes natural phenomena, other times just random information that would have escaped our ken were it not for the presence of chance. For me, Gup's essay proved the point itself, for I came upon it unintentionally on the back page of *The Chronicle of Education*—a publication I don't read regularly, but which I stumbled upon one day in an office in which I was working.

Randomness, serendipity, and surprise can account for some of the most pleasing aspects of life. A man goes looking for friends at a bar, steps into the wrong joint, and winds up meeting the woman with whom he spends his life. A college student in the library picks up an obscure medical journal and finds it so intriguing that she decides eventually to go to medical school and become a doctor. A wealthy matron watching the evening news sees a segment about dilapidated classrooms and decides to donate a generous sum to build a new school.

Still, accidental encounters stand in better stead in non-Western societies than in our own. While the Chinese, for example, celebrate chance in the I Ching, we tend to be suspicious of anything haphazard and unexplainable, and inclined to believe that the only good life is the meticulously planned one. We don't let fate carry us much.[2]

The control revolution reflects and reinforces this. We are becoming so accustomed to plotting our experience—what information we'll receive and what we'll filter out—that letting go induces a sense of nakedness. When we browse web sites or scan TV channels, it may seem like we are trawling for random encounters. But in a media world shaped by information overload and constant filtering, this meandering is likely to be increasingly rare. With thousands of channels to choose from, "surfing" aimlessly will probably be less and less common. (The Supreme Court has already observed that on the Internet "users seldom encounter . . . content accidentally."[3]) Rather, we will use sophisticated search-engine and agent technology to find just the content we want.

Personalization, to be sure, can help us in a variety of ways—to gain specialized knowledge, to explore our idiosyncratic desires, or just to clear a path through the data smog. But as appealing as it may be to tailor information and social interactions, too much personalization can diminish communal bonds, compromise democratic values, and deprive us of opportunities. Even Tim Berners-Lee, the inventor of the world wide web, wonders whether the technology he pioneered will allow people "to set up filters around themselves . . . and develop a pothole culture out of which they can't climb."[4]

To date, excess in these arenas has largely been blamed on direct marketers and others who have a financial interest in "breaking up America" into narrow niches.[5] But as opportunities for personalization expand, individuals will take on a greater share of responsibility for their information diet and their social encounters. Marketers may still try to feed us their narrow offerings, but increasingly we will each shape and be accountable for our experience.

Embracing Chaos

How, then, can we achieve balance with regard to personalization? To begin with, we must recognize the political nature of our filtering and personalization choices. We should also find trusted intermediaries who can bring us high-quality news of a *general* nature so that we remain broadly informed and engaged in our communities. Beyond being wary of specialization, though, we must occasionally give up control—even delegated control—and let ourselves be exposed to serendipity. Rather than trying to completely eliminate chaos, we should tame it and even embrace it somewhat. That way, we'll leave room for encounters with facts, ideas, and people that we might not otherwise know about.

> We must learn to say no to some forms of personal control.

It may, at first, be difficult to accept the idea that we should not always impose order on experience. Yet occasionally, the most appropriate response to technological empowerment may be what social critic Jacques Ellul called an "ethics of non-power." We must learn, Ellul says, to say no to some forms of personal control, to reject the idea that whatever we can do, we must do.[6] This notion of having power but not using it, or having it and giving it away, is essential to the possibility of reconciling individual and communal goals, private and public values. It should allow us to step back and think about the new capabilities we are acquiring and how we want to use them.[7]

If we are, for example, inclined to completely personalize our information or social environment, what does that really tell us? Is our resistance to giving up control that much different from the resistance to such divestiture that we have seen on the part of some powerful elites? Are we so bereft of confidence in our views that we need the constant reinforcement of a hypercustomized world?[8]

Some social critics even see glimpses of our obsession with control in such mundane activities as the way that we listen to music today. Gerald Marzorati argues that the digital music format, on compact disk and on the Net, encourages listeners to skip around so much that they may ultimately miss the real beauty of a musical recording. Listening to an album should be, Marzorati writes (and the emphases are his), "a matter of giving yourself over to *somebody else's* choices—*this* song, then *this* one, because it was conceived to be heard that way. The digital revolution promises precisely the opposite: you get to pick and choose, quickly, effortlessly, endlessly. What do *you* want, want right *now*?"[9]

As this critique suggests, much of our new technology seems to be used to defy complexity. The potential for oversteer, though, should show us that too much order may be as dangerous to public life and personal well-being as too much chaos. We should appreciate the benefits of allowing a bit of randomness to creep into our meticulously planned information domains.[10] We will have the advantage of perspectives other than our own.[11]

Students of a branch of science known as chaos theory might add that from such unstructured activity can come a richness of experience that never could have been planned—by us or by any trusted intermediary. Chaos theory suggests that there is often order in what appears to be random and complex—an "orderly disorder"—though it may be an order that will be difficult for us to grasp without an open mind.[12]

Open-mindedness, in fact, is as much a practical virtue as an abstract intellectual or moral quality. In the realm of politics, compare the plodding single-mindedness of the last rulers of communist Eastern Europe to the agile versatility of a Vaclav Havel or Nelson Mandela, both of whom were able to shift from being staunch dissidents to steadfast leaders. As historian Will Durant observed, "Great empires often bring about their own demise through excessive adherence to their own central principles."[13]

Similarly, the business person, professional, or student who is too narrowly focused will miss critical trends and rewarding opportunities. Indeed, in a labor market where the average employee has more jobs in a lifetime and a wider range of responsibilities during the course of each job, flexibility and breadth are assets for all workers.

Education and Encountering Difference

Schools and universities have a special role to play in seeing to it that citizens in the digital age do not fall into the trap of using individual control to indulge only their own interests. This is a natural fit for these institutions, because an open mind is the foundation of a liberal education. This should be the case from kindergarten—where a good teacher will always encourage students to try something new—to our most distinguished settings of higher learning.

The so-called culture wars of the last decade have dealt substantially with the question of what a student must know in order to be considered "educated." But participants in this debate, traditionalists and revisionists alike, have focused too much on what courses are required and what books are on reading lists, when what matters is the habits of mind we help our students to cultivate. A successful education is not born out of any fixed formula, but is achieved whenever we instill in a young person a real desire to learn for learning's sake. This state of broad curiosity suggests not only intellectual confidence but a willingness to be intellectually vulnerable. It requires not just "book knowledge" but exploration of personalities and environments that are unfamiliar. These encounters are essential to warding off the complacency and hubris that excessive personalization might bring.

Yale University president Richard Levin hit just the right note when explaining why he denied the request of a few religious students who wanted to live off-campus so they could avoid dorm-room activities they found objectionable. "This university has been committed to offering an encounter with difference as part of its educational mission," said Levin. "These students want the education, but they don't want the encounter."[14]

> The power that technology affords could be the most promising trait of the wired life or one of the most deceiving.

Encouraging people to encounter difference is another way of saying that we should occasionally give up control, loosening the taut order of habit and social constraint to see things from another perspective. The best teachers have always made this a central component of their instruction, whether through a great book, a history lesson, or a science experiment. Certainly there are others in our lives who have the ability to show us the value of openness: parents, friends, religious leaders, writers, artists. (And there are other institutions besides schools, including governments and corporations, that should see the worth in encouraging experiential breadth.) But educators have a unique opportunity to show young people why they should avoid too much privatization of experience. By the time they are adults, many individuals are so set in their ways, so accustomed to filtering out undesired voices, that they will not be receptive to hearing about, of all things, the need for open-mindedness. In other words, those who most need to hear the message of mental *glasnost* will probably block it out.

To achieve balance, then, society must be very careful about the message it sends to children about technology and individual control. They are the ones who can still be influenced before they have succumbed to endless opportunities for isolating personalization and other forms of oversteer—or, on the other hand, before they have learned to surrender to resistance of one type or another. Teachers, parents, and other adults need to explain to kids that, depending on how we handle the power that technology affords us, it could be the most promising trait of the wired life or one of the most deceiving.

One possible glitch with relying on schools and universities to combat oversteer is that these institutions themselves may be in jeopardy. Home-based "distance learning" online could become a popular alternative to traditional schooling for reasons of cost and convenience, and perhaps most of all because an online curriculum can easily be customized to the desires of parents and their kids.

Indeed, it is likely that those groups that oppose encounters with difference most—whether for religious, cultural, or socioeconomic reasons—would be the first ones to choose distance learning over traditional schools, from primary through higher education. Like news organizations and public officials, then, educational institutions face the prospect of being disintermediated—cast aside as inefficient and anachronistic. As in those other realms, we must recognize that schools and teachers (face-to-face teachers, that is) are truly unique and necessary intermediaries: they are community anchors. In fact, contrary to the recent American trend toward privatizing education with school vouchers, we must strengthen public education, recognizing that it is society's best opportunity to give children a sense of openness to difference, as well as a sense of common destiny and mutual obligation.

Speech, the Street Corner, and PublicNet

With its potential for individual empowerment and unfettered citizen interaction, the Internet has been a harbinger of a society in which citizens will engage one another in the vital conversations of a democracy. The Net won't fulfill this potential, though, unless it is characterized by a strong degree of diversity and fortuity.

Traditionally the legal institution that would have preserved these values is Paine's beloved public forum. What makes this communal space work is that no one there has total control—not government, corporations, or even individuals. Rather, there is a kind of orderly disorder.

As federal judge Stewart Dalzell wrote in his district court opinion in *ACLU v. Reno*, the Communications Decency Act case, "Just as the strength of the Internet is chaos, so the strength of our liberty depends upon the chaos and cacophony of the unfettered speech the First Amendment protects."[15] Judge Dalzell, though, presumed that in the absence of the CDA, the healthy chaos of free speech on the Internet would be sustained. He didn't consider the ways in which private actions might interfere with the openness and diversity of information, particularly since there are no legally mandated public forums online.

> The Net must have some of the serendipity of a Times Square.

There are, as I have noted, two distinct hazards that could get in the way: first, the restricted content options offered to us by gatekeepers like AOL or Microsoft; and second, the excessive personalization by individuals of information and social encounters. Both represent a poverty of choice and experience—in the first case, because we allow others to limit our options; in the second case, because we ourselves limit them, often unwittingly. (To be sure, the prospect of individuals, even a lot of them, ignoring one content source or another is not as troubling as a major gatekeeper doing so, because they don't have the concentrated power that the big intermediary does. But practically, as Paine's example showed us, the effect may be similar.)

This is not to say that we should never follow convenient links provided for us or customize content, only that we should be as open to accidents in this context as we traditionally have been in others. If communal interaction offline is increasingly (albeit gradually) being displaced by mediated life online, then the latter must be imbued with some of the chaotic vitality of the former. President Clinton has declared that the Net is becoming "our town square."[16] If so, then we should design and use the Net so that it has some of the serendipity of a Times Square or a Hyde Park Corner. The Supreme Court has said that any Internet user can be "a pamphleteer" or "a town crier with a voice that resonates farther than it could from any soapbox."[17] If so, then some balance between the unpopular speaker and the reluctant listener should exist online. Pamphleteers and town criers, after all, cannot be filtered out easily by reluctant audiences.

How do we achieve this open state of discourse online? Could we decide, for example, that powerful gatekeepers such as Microsoft should, like governments, have a special obligation to preserve the freedom of speech? Microsoft does not really censor anyone outright; it can't haul them off to prison or deny them a license to speak. But it does play favorites with content providers in a way that the U.S. Constitution would prohibit a state actor from doing.[18] Yet because the First Amendment only applies to state actors, Microsoft's screen bias cannot be challenged in court as a violation of free speech.[19]

There are, however, more creative ways to require Microsoft and other powerful access providers to save some space for voices that would not otherwise be heard, while also making sure that individuals remain open to the occasional accidental encounter that they would find in a chaotic community space like the public forum. All it should take, in fact, is a bit of enlightened tinkering with the code of the Net to:

- *Create an online space called PublicNet—a kind of street corner in cy-berspace*—where otherwise-invisible community groups, activists, and artists might occasionally have a limited but real opportunity to be seen or heard. Like the traditional street corner speaker, they would get an opportunity to confront their fellow citizens, who would in turn bear only a minimal burden to know that those speakers existed.

The challenge for PublicNet would be how to deal with all the potential speakers (thousands or more, potentially) who might want to confront their fellow citizens at once. The answer, though, is in the technology itself, and in serendipity.

First, the technology: PublicNet would appear first to the user as an icon either on the desktop or the browser. Clicking on that icon would take the user to an endless digital public forum, an aggregation of web sites and chat forums linked together voluntarily under the PublicNet umbrella. This collection would include not just ranting dissidents like Paine, but a vast array of enriching nonprofit sites dedicated to arts, politics, culture, and science (after all, much of this content is being displaced from forums like public access cable television channels).

Now, for the serendipity: The PublicNet icon on the desktop would have a randomness feature built into it, so that each user would see a different teaser from a speaker or group trying to get attention. These teasers would be just a few words that appeared on the relatively unobtrusive PublicNet icon. They might say "Save the Whales" or "I'm a Paine" or whatever. If the user chose to click on the PublicNet icon, she would be transported to that speaker's site. If not, within a few seconds, the message would change—just as if the user were a pedestrian walking through a crowded town square.[20] Once at a PublicNet site, the user could—again, as in a physical public forum—stay, go, or look more deliberately for certain content with the help of a PublicNet subject directory and search engine.

Some users (like some pedestrians) might try to ignore PublicNet altogether. But it's not hard to see how people might actually come to appreciate it. As on the street corner or public access cable, we might be exposed now and again to an unplanned encounter that we would find interesting and valuable. Indeed, we might begin to look to PublicNet to see what was going on there, the same way we might be conveniently tempted to click on a Disney or Time Warner icon on the desktop to see what is there.

Could the government require Microsoft or other large gatekeepers to place the PublicNet icon on their desktop? It shouldn't have to. A combination of corporate goodwill and consumer pressure should convince them that it's a good idea. But if Bill Gates isn't willing to give up the real estate, then the federal government should force him to do so.

Microsoft, to be sure, has a First Amendment right to speak in the manner it wants—or not to speak at all.[21] Yet powerful access providers (for example, cable television operators) don't get totally free rein in deciding what content they want to carry. If there is an important public interest at stake—and protecting community strength and the diversity of speech is such an interest, the Supreme Court has said—then a narrowly tailored mandatory access provision such as requiring a major gatekeeper to carry a PublicNet icon should be constitutional.[22]

PublicNet is also a fair re-creation of the traditional balance between the unpopular speaker and the reluctant listener, preserving some opportunity for confrontation while acknowledging the need for filtered order to protect one's sanctuary. Having to occasionally see PublicNet teasers floating by on one's computer screen is equivalent to, or less than, burdens we experience in other facets of our lives. A pedestrian may decline a leaflet from a canvasser, but he cannot erase his presence altogether. Similarly, a newspaper reader who reads only the sports section or the op-ed page can't avoid catching at least a glimpse of the other headlines as she flips the pages.

More than fifty years ago the Supreme Court noted that free speech would only be effective if speakers could find an audience. "The right of free speech," it noted, "is guaranteed every citizen that he may reach the minds of willing listeners and *to do so there must be opportunity to win their attention*."[23] In order for citizens to have their one bite at the apple—some minimal chance to engage other community members— we need something like PublicNet. It's a resource that could also prevent us from using the Net to narrow our horizons too much.

Embracing chaos, encountering difference, and building a serendipity-rich public discourse are not the answer to all our problems. Will poverty be lifted if the rich and poor can learn about one another? Will war end if the pacifist can wave a sign in the face of those who are indifferent to belligerence? No. But neither will democracy work if life is organized exclusively around the mantra "Where do I want to go today?"

If we aspire to some improved social condition, we could do worse than to start by creating accessible, even occasionally unavoidable, spaces in which citizens can confront one another and minds can be changed. It is not a magic bullet for society's ills, but it is better than isolation and silence. It also does not preclude the formation of more specialized forums dedicated to certain constituencies and issues, so long as participants in those forums do not abandon the idea of some minimal commitment to shared, open dialogue—particularly with members of their local communities.

Chapter 20

Surf Globally,
Network Locally

In the early 1980s, the New York Islanders, a hockey team based in suburban Nassau County, Long Island, won the world championship Stanley Cup four years in row. And each year they had a predicament. Without the equivalent of a Times Square or Pennsylvania Avenue, there was no place for the Islanders to hold a tickertape parade. In the end, the team's fans were left to gather outside the Nassau Coliseum while the Islanders drove around the parking lot.

This story, says urban critic Charles Lockwood, is emblematic of the problem of American suburbs, which often "have no heart, no central core, no sense of place that residents can lean on, take pride in or turn to in times of celebration."[1] Lockwood, however, sees a countertrend developing. Many suburbs are building Main Streets to create the sense of place and community that cities have. Ersatz as these thoroughfares may be, what they suggest is that even those who like the sprawl of the suburbs may be recognizing a need for the conviviality of an urban neighborhood.

This renewed interest in locality reflects a broader dissatisfaction in cities and suburbs alike, according to social scientists, with the isolation of contemporary life, the unraveling of family and communal bonds, and the numbing effect of consumerism.[2] Many observers likewise believe that the rapid growth of the Net as a communications tool is a sign of people's desire for a more connected way of living. The growth of online communities, in particular, suggests that there is a deep unfulfilled longing among citizens today for affiliation with others.

In the view of cyber-romantics, that hunger can be satisfied by the new individual control that technology provides. The ability to tailor social environments and convene with likeminded others, they say, is the perfect antidote to social isolation. This idea, in fact, goes back to the origins of the Internet. In the 1960s, J. C. R. Licklider, a pioneer of networked communication, wrote that "life will be happier for the on-line individual because the people with whom one interacts most strongly will be selected more by commonality of interests and goals than by accidents of proximity."[3]

> We must recognize, for selfish and societal reasons alike, the importance of focusing on the local.

The truth, though, is more complicated. The ease of exit from online associations means that we may form weak bonds with others faraway, often at the expense of the strong ties that come from shared experience with the friends and neighbors who live near us. Local communities, and even nation-states, may fragment and weaken as a result. The neo-Luddite answer to this dilemma often is that we should reject technology altogether, returning to our bucolic roots.

A more balanced approach, by contrast, calls for a reconciling of personal needs and communal obligations in a digital world. On the one hand, this means acknowledging the sometimes exhilarating adventure of indulging oneself online. No one can deny the wonderment and value of being able to form relationships with far-flung others based solely on common interests. At the same time, it means not having illusions about the durability of those bonds or their ability to satisfy fully our deepest needs.

We must recognize, for selfish and societal reasons alike, the importance of focusing on the local. For most of us, this is where we will find a true sense of belonging; shared experience, even if not ideal, creates a sense of commitment. For all of us, this is where democracy and social justice must first be achieved; getting our own house in order is always the first priority. We must therefore use the Net as a vehicle not just for occasional escapism, but to enhance local engagement online and off.

Community Networks

Efforts to employ technology to strengthen local communities are not new. They have been tried since the dawn of cable television in the 1970s and, for over sixty years, via community radio programming.[4] Those communications technologies, though, are one-to-many. What makes the Net so promising as a tool of localism is its interactivity, as well as its flexible and nearly unlimited capacity.

Many early Internet enthusiasts have been strong supporters of "community networking," an approach that encourages locally based online communication, often at no charge to users. Community networking has its origins in services such as the Free-Nets, which emerged in the 1980s and early 1990s to offer online access, sometimes along with local news and information. Most were noncommercial, with no advertising and no subscription charges. Often they were text-based bulletin board systems—and were not easy for novices to use.

A good share of these early services, in addition, were not so much *about* local affairs as they were a way to get online for free.[5] As a result, Free-Nets and other community networks suffered as America Online and other inexpensive (and more alluring) gateways to the Net became available. By the late 1990s, many had gone out of business, as had the National Public Telecommunications Network, an umbrella group of Free-Nets that was founded in 1986.[6]

Still, more than a hundred Internet-based community networks in the U.S. have continued to thrive, such as Charlotte's Web in Charlotte, North Carolina; Liberty Net in Philadelphia; the Seattle Community Network; and Blacksburg Electronic Village in Blacksburg, Virginia.[7] Founded in 1991, Blacksburg Electronic Village has been one of the more successful of these endeavors. It counts more than 60 percent of Blacksburg's citizens as participants, in part because they can access the network for free. Senior citizens chat with their neighbors online. Parents keep abreast of what their kids are doing in school and exchange email with teachers. Citizens use web-based surveys to communicate with their municipal government about spending priorities.

A key feature of successful community networks, in fact, is the opportunity they provide citizens to talk—with civic leaders and one another. Users don't just want information fed to them, they want to generate conversation themselves.

In a community network in Amsterdam, for example, citizens talk about keeping the city's largest park in shape, they argue about Amsterdam's proposed transformation from city to province, and they bombard politicians with questions about Holland's abstruse tax laws.[8]

Similar results were apparent even in a short-term case study involving a group of London neighbors. Microsoft gave them computers, Internet access, and a way to communicate with one another online. Participants used the technology to exchange information about local services. Kids asked questions about homework. There was a debate about a proposed change in local parking rules, and some members even organized to do something about disruptive vibrations from a nearby railroad. The dialogue, moreover, appeared to translate into stronger ties among neighbors. "I used to know maybe 5 or 6 people in the street, now I know at least 40 of them quite well, and some very closely" one participant said.[9]

Even some early online services that didn't start as community networks appear to have succeeded precisely because members were located mostly in one geographic area. I mentioned one of the best known of these, the San Francisco–based Well, in the earlier discussion of the *Time* cyberporn episode. Though the Well was never intended to be about San Francisco or just for people from San Francisco, its founders knew from the start that a sense of local culture would be an important component of the online community.[10] Most interestingly, perhaps, they recognized the value that regular face-to-face contact would have for members. Monthly Well parties were therefore instituted in the San Francisco area and became an important element of the online community's identity.

Opportunities for face-to-face interaction seem, in fact, to be another essential component of successful online communities. Howard Rheingold, an author and Well member, wrote in *The Virtual Community* that a pivotal experience in the evolution of the group was the suicide of a member who had shared much of his depression online.

After attending the funeral, Rheingold wondered: "How could any of us who looked each other in the eye that afternoon in the funeral home deny that the bonds between us were growing into something real?"[11]

Though Rheingold never expressly mentions face-to-face contact as a requirement for success, most of the vibrant virtual communities he describes relied on such interactions. Thus, Rheingold notes that as members of one of Japan's first virtual communities, COARA, began to meet in person on a monthly basis, they "began to think of themselves as a community."[12] And CIX, a London-based online community, underwent what Rheingold calls "a familiar evolutionary cycle: disparate characters meet online, find that they can discover depths of communications and deep personal disclosures with each other online, form equally intense friendships offline . . . ."[13]

Similarly, Echo, a prominent New York–based online community, offers regular events such as readings, a film series, bar gatherings, and softball games. As Echo's mission statement says, "We know that the best online communities are never strictly virtual."[14]

Contrary to the utopian notion that the Internet will lift us above the confines of geography, then, the history of online communities suggests that people want to convene with their geographic neighbors, both online and off.

Local Gateways

Given this fact and the success of some community networks, it might seem that little needs to be done to achieve balance between our desire to surf globally and our need to network locally. Yet as the control revolution presents the possibility of a more alluring universe of distractions and greater social isolation, our emphasis on localism must become stronger and more explicit. We need to:

- *Build high-quality, web-based local community networks* that are ubiquitous, accessible, and interesting enough so that all Net users will want to use them, at least some of the time.

This would ensure a degree of involvement with community issues and engagement with actual neighbors. These networks should not be final destinations, though. Instead, reflecting the idea of a local/global balance, we would do better to think of them as local gateways to the global Net—and to offline interaction as well.

Like entry ramps, these gateways should allow users to go anywhere. Yet, learning from the successes and failures of predecessors, they must provide stimulating content about local issues and, even more, an opportunity for users to talk with one another. There should be resources and discussion about issues that people really care about: recreation and entertainment, sports teams, politics, schools, shopping, crime and safety, and consumer assistance. This alone should entice people to visit.[15] And as local gateways facilitate dialogue among community members, eventually empathy, interdependence, and cooperative action will follow.

For users without Internet access, the local gateway could be the service they use to get online—for free. The government could thereby kill two birds with one stone: working toward both universal access and localism. Following the lead of existing community networks, Internet terminals could be put in schools and libraries, churches, public housing projects, and recreation centers.[16] For those who already have online access, the local gateway could be used as a portal site on the web.

The architecture of the local gateway is crucial. Its blueprint should be influenced not just by a local/global balance but by the other balances I have mentioned, between convenience and choice, power and delegation, order and chaos. For example, the ability to speak and be heard should not depend solely on one's resources and there should be ample opportunities for serendipitous encounters. This online "commons" must be a worthy complement to the physical public commons—not a substitute, but an extension.[17] It should thus have all the quirks and flavor of the geographic community for which it is a digital annex.

In terms of content and design, there are two obvious models for the kind of local gateway I am proposing. One is existing community networks, which are generally superb examples because they empha-size localism and dialogue. Sometimes, though, community networks are an end in themselves, instead of an entrance to the whole Net. To draw a larger audience, the gateway format is better because it becomes a routine starting place for users, while not confining them.

The opposition by some community networks to partnering with business may also be counterproductive. Blacksburg Electronic Vil-lage, for one, claims to have benefited greatly from the fact that it is a partnership among government (the town of Blacksburg), academia (Virginia Tech), and industry (Bell Atlantic, the local phone company). More than two-thirds of local businesses are on the Blacksburg net-work. This provides a convenience for users and a commercial oppor-tunity for local vendors who might otherwise lose business online to huge cyberstores based outside the community.

At the same time, local gateways should not be overly commercial-ized. In particular, citizens should shun attempts by corporations to fabricate communities just so they can use members as a target audi-ence for sales and advertising. It's a practice that has been tried on the web, though fortunately with little success so far.[18] Businesses would be better off working in cooperation with community groups and lo-cal governments. And citizens should welcome their participation, par-ticularly if they are locally based, and so long as they maintain a civic-minded spirit. In fact, the damaging effects that electronic com-merce could have on local economies (on stores such as bookstores, for example) could be offset, to a degree, by the ability of community members to patronize online versions of their favorite neighborhood stores.

An unlikely boost for local gateways might also come from city-ori-ented commercial web services such as those provided by City Search, Yahoo, Microsoft's Sidewalk, and AOL's Digital Cities. Some Ameri-can cities have as many as half a dozen of these sites competing for the public's attention.[19] With their collection of information such as local news and weather and services such as free email, these sites provide a second model for local gateways. Community networking activists have traditionally seen them as the enemy—because of their com-mercialism and the fact that they attract individuals away from nonprofit sites. Yet under the right circumstances, these sites could help anchor individuals in their communities. They could become partners in the formation of local gateways.

For this to happen, citizens need to leverage the power that interactive technology gives them. We need to organize and tell these city-based portals that to win our attention they must give something back. They must, for

> The goal of the control revolution should not be to replicate the world we know, but to improve it.

example, donate substantial online resources—such as free web site hosting and design, chat forums, dial-up access, and hardware—to community-based organizations such as tenant groups, parent-teacher associations, charitable entities, activist groups, and religious institutions. They must offer Internet authoring tools that anyone can use to create their own dialogue forum. And they must find people to lead moderated discussions and otherwise work to strengthen the online presence of local communities.[20] Moreover, all of these resources could be coordinated with PublicNet so that local groups and individuals have a chance to reach out even to those who don't regularly use local gateways.[21]

Finally, local gateways should not be seen as a panacea for community activism. They must instead be part of a larger strategy of face-to-face local engagement—which may nonetheless be more effective and more enjoyable thanks to local online interaction.

The opportunity for truly open and effective community conversation will depend on whether our digital tools are primed, in design and use, for online escapism and total filtering, or whether we balance personalization with other values, such as broad exposure, social awareness, and community strength. If we fail to do so, the problem won't just be that we are all occupying different chat rooms or reading different news online. The insularity of cyber-experience could become the insularity of experience, period.[22]

Steam and rail gave us the opportunity to flee far from our places of birth; telegraph and telephone allowed us to conduct our business and social lives from a distance; television drew us away from communal spaces and gave entertainment and news a monochromatic gloss. The goal of the control revolution, if it can be said to have one, should not be to replicate the world we know, but to improve it. We should therefore use the Net to explore the farthest reaches and to fortify the communities in which we dwell.

If we do, we may find that community is the ultimate safeguard of balance—a counterweight to both institutional power and individual power, a balm against resistance and oversteer.

Chapter 21

The Tools
of Democracy

A decade and a half ago, Ithiel de Sola Pool published *Technologies of Freedom*, a highly influential history of state responses to different communications technologies. After comparing the regulatory environments that have governed print and post, telegraph and telephone, radio, television, cable, and emerging computer networks, Pool concluded that liberty and progress would be best secured if government did not regulate communications technology. He believed that new media would foster freedom, as books had for centuries, if the state just left them alone.

Pool has consequently been seen by many as a forefather of today's libertarian consensus on communications policy. Yet his call for government restraint was a product of the times—a response to the cold war legacy of state manipulation of media in communist countries, as well as some misguided government interventions in the U.S. and other Western democracies.

At the turn of the century, the situation has changed significantly. Market capitalism and laissez-faire are now the guiding lights of modern economies generally and of communications policy in particular. As telephone and broadcast networks throughout the world are privatized and deregulated, and as technology evolves at a faster-than-light pace, control of communications resources has shifted almost completely from the public sector to the private sector. Societies are relying on markets, moreover, not just to distribute goods but to solve social problems.

This is a mixed blessing for individuals. On the one hand, the combined power of technology and the market can give us an intoxicating sense of personal dominion over our circumstances. Often it may seem that we can rely on these twin engines to do anything. But the control revolution, as I have emphasized, may unfold in unpredictable ways. The apparent boon of an unfettered market can be a mirage that masks the dwindling of democratic values and real personal autonomy. Tempting as it may be to rely on the private sector to regulate itself and on consumers to protect their own interests, doing so exclusively will cause core public goods like free speech, privacy, equality, and community to be relegated to the status of commodities.

Part of the problem is that the control revolution does not assure a linear transfer of power from institutions to individuals. Most notably, if laissez-faire deregulatory policies remain dominant at the same time that state authority diminishes, corporations—particularly large multinational conglomerates—may take advantage of the resulting power vacuum. The new individual control might empower consumers at the same time. But given the hugely disproportionate power of corporations, we simply may not be able to leverage our new abilities against the private sector. Without some help, the control revolution could come to a sputtering halt.

In the last five chapters, I have argued that balance can be achieved if we apply principles in context, reconcile convenience and choice, and recognize the value of middlemen, accidents, and localism. These admonitions have focused a fair amount on what individuals can do to make the revolution come out right—recognizing, as Esther Dyson says, that in a world of diminishing state power "individuals will need to assume greater responsibility for their own actions and for the world they are creating."[1]

There are many areas in which just ends can only be attained if we act together through government.

Yet I have also been trying to show that individual action on its own is not enough. To attain balance, we also need collective action, often through government intervention. Thus, a solution to protecting kids from indecency may require some state regulation of commercial middlemen. Courts might have to intervene to make sure that trusted systems do not undermine fair use. A competitive software market may require stimulus from government. And the state may have to regulate or provide incentives for powerful gatekeepers to create a public forum online or to encourage localism.

Continuing this progression, my final axis of balance emphasizes more explicitly the crucial role of collective public action. There are many areas in which just ends cannot be attained by individuals on their own. We must act together to promote equality and opportunity and to protect fundamental rights.

We need, in other words, a balance between the market and government, between unimpeded choice and democratic values.[2] Some of this governance might be done in novel ways by citizens and organizations and even industry.[3] Not all governance requires the action of the state; this is particularly true given the opportunities for self-ordering that new technology provides. But traditional government intervention will also be necessary. As even libertarian icon Friedrich Hayek wrote: "In no system that could be rationally defended would the state just do nothing."[4] The goal is to see that the Net and other new media are not just technologies of freedom, but tools of democracy.

Markets and Safety Nets
(the Case of Privacy)

Despite the claims of cyber-romantics that the Net is a sovereign territory unto itself, the idea that the government has a role to play in the online world is not a remarkable proposition: the U.S. government sponsored the development of the Net; it would make little sense for it to abdicate completely, particularly when it comes to protecting consumers.

Generally, the Clinton administration has seemed to grasp this, but its Internet-related policies sometimes have gone in different directions in rhetoric and practice—directions that can seem contradictory.

For example, even as the president has argued that government should not "stand in the way" of the growth of the Internet,[5] his administration has sometimes done just that when it comes to encryption technology, the free flow of information, and protecting the public domain. At the opposite extreme, the administration has sometimes been shortsighted in its insistence on a hands-off stance, failing (at least initially) to recognize where individual control needs to be supplemented by collective action.

An instructive example of this has been the administration's approach to protection of consumer privacy. In keeping with its emphasis on industry self-regulation, the administration has supported the development of a "market for privacy" in which individuals would bargain with vendors over data collection practices. This approach, as I noted in Chapter 15, is fraught with problems that demonstrate why protection of important values like privacy shouldn't be left solely to individuals—and, more generally, why balance between the market and regulation is necessary.

Individual control of personal information can give us a strong degree of sanctuary. We can determine our own privacy preferences, benefit from trading certain data about ourselves, and expect to be more informed about how our information will be used. But the market for privacy has failure points and blind spots that should convince us that it doesn't make sense to bypass government altogether when it comes to protecting our data. Rather, we should rely on government as a trusted intermediary that can set a baseline of privacy protection. We need it to:

- *Establish a privacy safety net*, a minimal level of personal information privacy that cannot be taken from us even if we think that bartering it away with a click of the mouse makes sense. At the very least, it should prevent vendors from collecting information from kids and place strict limitations on the collection of medical and financial information.

Encouraging steps in these directions are being made. In late 1998, Congress authorized the Federal Trade Commission to develop rules requiring web sites to get permission from a parent or guardian before collecting information from a child twelve years old or under.[6] At the same time, the commission reiterated a longstanding warning to the online industry that it would recommend passage of further legislation to protect privacy unless strict industry standards for collection and use of personal information were adopted and observed. President Clinton has also called for the passage of legislation to protect medical privacy.

Generally, the federal government should do more to create an environment where personal information privacy is the norm, not the exception. The burden should be on data collectors to justify their practices rather than on hapless individuals trying to protect themselves in a perplexing new marketplace. As it is now, we're at a disadvantage.

By contrast, the European Union's privacy directive requires data collectors to get explicit permission from individuals before gathering sensitive data. It provides individuals with a right to review, and correct errors in, collected information. And it gives them an express remedy in the case of violations.[7] We would do well here to follow the model of the Europeans, who believe that though the market and technology can be used to enhance privacy, government must do its part.[8]

In addition to establishing a privacy safety net, government action is needed simply to provide a concrete legal regime in which a market for privacy can work efficiently and equitably. To protect against data theft and insure fair dealing and compliance, for example, an enforcement mechanism is necessary. Whether or not we trust the self-regulatory efforts of industry groups like the Direct Marketing Association, they surely can't control the actions of fly-by-night companies that swoop down on the Net to pick up the trail of an unsuspecting mouse. Many European nations have privacy commissioners or other central authorities that are responsible for privacy protection. The U.S. government should follow suit and:

- *Create a federal agency to coordinate privacy protection* domestically and abroad, or at a minimum designate an existing agency as the clearinghouse for all privacy-related policy matters.

Code and Commerce

In addition to providing safety nets where markets fail, government needs to protect consumers in a changing commercial world. How, for example, will we preserve those public functions—such as taxation, safety, and disclosure—that have relied on the presence of middlemen who may be disappearing? The answer is in the code of the Net. Recognizing again the value of applying principles in context, we need to ask government to:

- *Support or create code intermediaries* that can assume some of the public-interest functions once undertaken by traditional middlemen.

These new intermediaries might be somewhat like Amazon and the other digital go-betweens that are reintermediating markets. Only now the new value added would not be oriented toward price or choice, but toward protecting the public good.

In the case of sales taxes and the prospect of lost revenue, for example, we might find a technical solution that taps into the new architecture of electronic commerce. Commercial transactions will increasingly rely on digital money, sometimes called e-cash, that can be transferred across the Internet swiftly and securely. To ensure the security of e-cash—that it's coming from the right source and hasn't been tampered with, and so on—third parties, not unlike credit-card companies, are beginning to process transactions and authenticate the identity of users.[9] The government could take advantage of this situation, requiring these transactional intermediaries to automatically remit a small portion of the paid amount to the proper taxing authority.

In fact, the problem of competing jurisdictions claiming a share of an online sales transaction might even be solved through use of these digital tax collectors. It would not be hard, say, to split six cents on the dollar among three different states representing the location, respectively, of the buyer, the seller, and the computer server that processed the transaction.[10] Unpopular as online taxation may be today, if local

governments experience a substantial decrease in revenue and are forced
to make cuts to basic services like education and crime-fighting, there
will be strong popular support for digitally enabled tax collection.

In some areas, we might also ask government to stimulate the development of code intermediaries that do not yet exist. To prevent
fraud and encourage information disclosure, for example, we could
draw upon the interactivity and flexibility of the Net to create intermediaries that would perform some of the same tasks as traditional
brokers, agents, and dealers. A digital seller of securities could be required to place a prominent link on its web site to a code intermediary—an independent third party, such as the National Association of
Securities Dealers—that would appraise an offering or report whether
the seller had any record of misconduct or fraud.[11] A seller might even
be required to post an independent rating or seal of approval from this
middleman directly on its site. This is not unlike requiring federally
insured banks to post a sign that says "FDIC insured"[12]—the absence
of such information can sometimes be a more valuable indicator for
consumers than its presence. Similar attempts in other contexts to use
information disclosure as a form of indirect regulation have proved
fairly successful.[13]

There will be objections to government playing cyberspace architect in this manner. Libertarians will argue that the state should not
interrupt the self-ordering of the market. And if industry can succeed
in creating these intermediaries on its own, perhaps an appropriate
scheme of governance can be achieved without much intervention by
the state. Yet self-regulation, as the case of privacy protection suggests,
will never be completely sufficient. Government will have to lay foundations for commerce, and it will have to keep an eye on the emerging code of the Net to ensure that the public interest is protected.

At the same time, as we've seen in the areas of encryption and free
speech, attempts by government to influence that code can have dangerous results as well as favorable ones. Not only are there potential
problems of legitimacy and jurisdiction due to the Net's relatively borderless nature, but it is difficult for the state to operate in a domain
that is so fast-changing and technically complex. Additionally, we are
unaccustomed to applying traditional checks on government power,

constitutional and otherwise, in the realm of obscure high-tech protocols. Though government has a role to play here, citizens have an equally important role in terms of watching to see that the state does not abuse its power or even wield it incompetently. The challenge we face is to strike the right balance—preserving certain functions of intermediaries when the public interest requires it, while allowing the market to cast others aside.

Checking Power

Control of communications resources is another critical area in which balance between the market and government is needed. Without hampering the growth and efficiency of emerging industries, we need effective ways to check concentrations of power in this realm, because it touches so closely on knowledge, public discourse, and freedom of speech. The promise of individual empowerment will likely be dashed if a few large companies have unrestricted control of our emerging communications infrastructure.

The supposition in recent years has been that as new technology does away with the scarcity of the airwaves and the costs of large-scale printing, anyone will be able to express themselves and otherwise take advantage of the bounty of new technology—so long as they are online. Focusing on this contingency, public-interest initiatives have mostly been concerned with establishing universal access to the Net. Federal programs such as the "e-rate" plan, for example, subsidize Internet access for schools and libraries in disadvantaged areas. And many impressive volunteer efforts have been undertaken to put computers and Internet connections in classrooms.

Online access is, of course, a necessary first step, but it is not enough to ensure that the full democratic potential of the Net can be achieved. Large gatekeepers, for example, may influence, or even restrict, the ability of users to speak freely or to select their own content options. The features on Microsoft's Windows desktop that steer users directly to its preferred content and commerce sites would appear, as the Justice Department's antitrust suit alleged, to be a classic case of a dominant access provider giving an unfair boost to itself at the expense of others. Even Pool, the champion of laissez-faire, insisted that the "monopolist of the conduit not have control over content."[14]

We should, as a result, think again about how to apply a timeless principle in a changed context. The principle is not complex:

- *In a democratic society, those who control access to information have a responsibility to support the public interest.* By dint of their power over such an important resource, these gatekeepers must assume an obligation as trustees of the greater good. Indeed, barring some clear showing that they are bearing this burden voluntarily, government should impose it upon them.

Simple as this public-trustee idea may be, finding the right rule to make it come to life in the context of the emerging Net may not be so easy. In other media contexts, analogous schemes have been fairly straightforward. Major newspapers meet their public obligations on their own initiative by covering elections and public affairs, by running public-service advertisements, and by turning over some space to letters-to-the-editor and opinion pieces. Broadcasters engage in similar voluntary efforts and, notably, are subject to laws requiring them to dedicate a certain amount of programming to governmental, educational, and citizen use. In both cases, the idea is that these information gatekeepers should share the *audience* they are able to get because of their commercial power with those who deserve attention, but can't afford it.

At least that's what is supposed to happen. In practice, newspapers and particularly broadcasters often don't do enough for the public. The increasing consolidation of corporate newspaper ownership in recent years has led news staffs to be cut and coverage to be simplified. And television broadcasters have generally tried to get away with the bare minimum in terms of providing children's educational programming, airtime for political candidates, and public affairs coverage generally. With the congressional giveaway of the digital television spectrum, valued by some analysts at as much as $70 billion, there has been an even louder outcry that broadcasters owe more to the public interest.

Whether or not it is fulfilled, the underlying idea, though, is the same: Major access providers are obliged to give something back to the public, to the community that sustains them. Applying this principle in the context of our new media environment requires some tinkering. As Pool rightly predicted, mandating broadcast-style public access requirements may not work in a world without channel scarcity. Indeed, under current legal precedents, it may not be constitutional. Laissez-faire proponents therefore conclude that online gatekeepers should be as unhindered as newspaper publishers. Yet that analogy doesn't work either because online gatekeepers have the potential to be much more powerful conduits—to other content, to commercial activity, and to all types of communications—than any newspaper or magazine.

Ensuring that these large access providers don't utilize their power unfairly must therefore be a societal priority. And though there are steps individuals and nonprofit organizations can take, this is an area in which we need the state to act for us as a countervailing force to powerful commercial gatekeepers. Drawing on some ideas I have already mentioned, here are a number of ways in which government might act. It can:

- *Limit the intellectual property protection* of operating systems and other software that succeeds in dominating a market and effectively locking out viable competition. The success of Microsoft Windows rests on the product's copyright, yet that right is a state-granted benefit that need not be taken for granted. At its core, copyright is about protecting creative works, not utilitarian products—particularly not those that, because of network effects, become monopolies. If the social cost of such an extended grant of power to one company is inconsistent with the public interest, the right thing to do is limit it. At a certain point, then, the code of a dominant product (like Windows) would enter the public domain and competitors could build on it to offer the public alternative products. This solution may, indeed, be preferable to protracted antitrust lawsuits.[15]

- *Pursue cases in the courts of law and public opinion.* In the Microsoft case, Justice Department officials proved that government can hold companies accountable for anticompetitive behavior and, at the same time, raise public awareness about the politics of choice in the digital age. This includes provoking debate about what constitutes fair play in the complex and fast-changing world of new media.

- *Promote diverse and broad-minded use of interactive technology* by, among other things, encouraging or (if necessary) requiring powerful gatekeepers to give their audiences easy access to something like PublicNet: an online public forum where the views of community groups, nonprofits, and individuals could be heard.

- *Stimulate a competitive software market* by funding research and development of nonproprietary open-source technologies.

- *Support local gateways* or other community networks that have a firm commitment to providing space for citizen dialogue. As in Blacksburg, local governments can lead the way in establishing partnerships among community activists, educators, and local businesses. Government should also encourage commercial online services to recognize their obligation to support local communities, using financial incentives if necessary.

In each case, the aim would simply be to preserve the longstanding idea that dominant media entities have a duty to act, at least some of the time, as trustees for the public interest.[16]

New Governance

The sublime myth that online interaction occurs on some uncharted, lawless frontier has lost its luster. Thoughtful lawmakers and policy analysts are recognizing that government has an important role to play in this realm. Looking at their own legislative agenda, it would be difficult for lawmakers to conclude otherwise. In the 105th Congress, which ended in 1998, more than one hundred bills related to the Internet were proposed, on topics including content control, privacy, intellectual property, the Y2K problem, infrastructure, and taxation of electronic commerce.[17] A similar growth in legislative activity occurred in state legislatures and at the international level in treaty organizations such as the World Trade Organization.

Not all of this activity, of course, should be applauded. When lawmakers approach the Net, they often do more damage than good. In part, this is because of the novelty of the technology—and particularly government's lack of familiarity with regulating the code of the Net. Its attempts to alter code often seem to affect individual behavior to a greater degree, and in different ways, than it intends. Also, lawmakers sometimes fail to see the need to map existing principles to new contexts rather than starting from scratch. Or they apply old rules without accounting for the new context of the Net.

These missteps should become less common, though, as the Net becomes more familiar and indistinguishable from the rest of our communications landscape, and as advocates succeed in educating politicians. In some circumstances, the digital vanguard would do well to take a more sympathetic approach to the concerns of government. In order for public officials to appreciate the benefits of the control revolution, they must be shown that this transformation is not an invitation to anarchy or a blunt rejection of all we know. In fact, lawmakers should know that, without their help, the control revolution won't be resolved in a manner that is consistent with individual well-being and democratic values.

An Internet policy report released late in 1998 by the Clinton administration suggested a growing awareness that government's role in the online realm is not so different from its role in the lives of citizens generally. This was an improvement on the rhetoric of a year earlier, when Clinton called the Net "a global free-trade zone."[18] It was, in fact, more of a return to the measured language of a 1993 Clinton plan, which stressed that there were "essential roles" for government to play in developing new media and ensuring that it would benefit all Americans.[19]

Generally, advocacy groups and concerned citizens alike will have to become more attuned to the contours of Internet policy—particularly code regulation—for this is where many future debates about social welfare, civil liberties, and economic justice will take place. High technology will not be a specialized domain for computer scientists and engineers. It will be an area that affects all of our lives intimately.

Finally, when it comes to balancing the market and government, we should recall that changes are occurring in terms of how and where governance takes place. In part this is because of the globalization that technology makes possible. But it is also because private entities, including corporations and nonprofit standard-setting organizations, will be making many quasi-governmental decisions about technology that will shape our communications landscape and hence our world.[20]

This can be advantageous, in the sense that these alternatives to traditional governance may produce more efficient results than bureaucracies could. We must, however, be attentive to the power of this new code regulation. We should scrutinize it, judge it by democratic values, and demand that it be accountable. Otherwise we'll be facing the prospect of the privatization of public policy.

Equally troubling is the way that the social contract envisioned by John Locke may go from metaphor to reality, as some cyber-romantics literally urge citizens to cast aside the state and shop for the private "government" that suits their needs. As James Dale Davidson and Lord William Rees-Mogg write approvingly, "you will be able to obtain governance at least as well customized to meet your personal needs and tastes as blue jeans."[21] For the right amount of money, in other words, you can buy security and infrastructure—and no obligations to others. This is, of course, the kind of solipsistic extremism that I have been arguing against. It is oversteer to the max.

What might happen in such a world? Traditional government figures would go from being resented and mistrusted to jettisoned and ultimately irrelevant. Even if it seems on occasion that this is what some of them deserve, we can't let the weakness of our actual circumstances cause us to lose all faith in what is possible. Such a flouting of obligation and authority, moreover, would jeopardize freedom. Could American democracy survive if its political institutions—the presidency, the Congress, the Supreme Court—were unable to establish and maintain authority? What about schools, the press, or religious institutions?

As political philosopher Jean Bethke Elshtain says, "Authority and liberty are not opposites; they are twins. If we kill one, we lose the other."[22]

From a more balanced standpoint, then, the control revolution confers upon us remarkable opportunities, but it also brings new burdens. Individuals *will* be more responsible for upholding freedom, safeguarding democracy, and creating a civil society—and this itself is changing the nature of governance. "More than ever before," says British author Charles Handy, "we are on our own, left to forge our own destinies."[23]

Yet all the individual control in the world will not build a just society or make our own lives whole. It will, in fact, lead us far from these goals. Our new individual power must be balanced by cooperation, mutual obligation, and collective action. Only then will we be able to enjoy its benefits.

In this pursuit, we must not underestimate the unique contribution that the state can make. It can intervene, in some of the ways suggested above, to protect individual freedom, power, and rights. And, more broadly, just as it promotes balance in areas like personal health, savings, and environmental protection, it can do the important job of promoting balance with regard to the complex ecology of control that we are beginning to encounter.

On May 21, 1944, Learned Hand, one of the great American judges of the twentieth century, addressed a group of more than 100,000 recent immigrants at a naturalization ceremony in New York City's Central Park. As war raged on in Europe, it was an especially poignant time to become an American citizen. Additionally, with more than a million people reportedly in attendance, Hand was addressing one of the largest crowds ever assembled in the U.S. It is not surprising, then, that he took the opportunity to reflect on the meaning of freedom and the role of political and legal institutions in securing it.

Yet Hand's comments were hardly predictable. Rather than heaping praise on the U.S. Constitution, the Congress, and the courts, Hand told his audience to look within for the real source of freedom.

"Liberty lies in the hearts of men and women," he said. "When it dies there, no constitution, no law, no court can save it; no constitution, no law, no court can even do much to help it."

On one level, Hand may have been preaching a bit of old-fashioned American self-reliance. But his message was more complex. He cautioned also that individual power, like state power, could go awry. Real freedom, he said, requires a measure of restraint and a healthy sense of limits.

"It is not the ruthless, the unbridled will; it is not freedom to do as one likes. That is the denial of liberty, and leads straight to its overthrow." He continued:

> A society in which men recognize no check upon their freedom soon becomes a society where freedom is the possession of only a savage few; as we have learned to our sorrow.

> What then is the spirit of liberty? . . . The spirit of liberty is the
> spirit which is not too sure that it is right; the spirit of liberty is the
> spirit which seeks to understand the minds of other men and women;
> the spirit of liberty is the spirit which weights their interests along-
> side its own without bias.[1]

Apropos as Hand's comments were half a century ago, the vision of freedom he was sketching is, as I have suggested here, even more relevant today given the new power that we have at our disposal.

There are two complementary components to that notion of freedom. The first component identifies the individual as the source of liberty. It acknowledges the importance of traditional institutions—such as representative democracy, local communities, and an independent press—but emphasizes that the flowering of a civil society depends most on the personal convictions and commitment of citizens.

Governments, corporations, and other institutional powers must therefore allow individual freedom to flourish. They must recognize that technologies such as the Net give individuals the ability to make decisions that they never could have made before. Individuals themselves, in turn, must rise to the challenge of increasingly being responsible for their own well-being and that of society at large.

That responsibility, in fact, leads to Hand's second component of freedom, which is a rather more delicate one: In order for individuals to achieve real liberty and justice, they must recognize limits on their autonomy—for reasons of both self-interest and the greater good. This may mean forbearing from some action, deferring to the judgment of others, and sharing power.

What holds these two components of freedom together, I have tried to show, is the notion of balance. Even as public and private authorities must take into account the new personal control, there must be equilibrium between individual and institutional competence, personal goals and common ends, self-interest and public interest. Dedicating ourselves to such a balance of power means that we will take advantage of our new abilities, but not abuse them—that we will both exercise our rights and fulfill our obligations.

This is no easy task. Max Frisch, a Swiss playwright, once described technology as "the knack of so arranging the world that we don't have to experience it."[2] If there is anything we can be sure of, though, it's that the control revolution will require us to experience the world head-on and full tilt. Resistance to the new individual control may be fierce and difficult to overcome. Achieving real choice will be more taxing than relying on simplified options. And it will be hard to slow the galloping momentum of change and prevent oversteer. But if thresholds can be obtained and limits set, then the control revolution can thrive, the existing order can be reformed, and the revolution can come to rest—overcoming both opposition to change and those who would push change to dangerous extremes.

For individuals and institutional elites alike, this will entail adopting an ideal of empowerment via technology that is based on something more than just accomplishing—or thwarting—everything that these tools make possible.

For our own sake, we need to see that living well in the digital age means more than just having complete dominion over life's decisions. Personal freedom requires knowing when to relinquish authority, either to chance or to the wisdom of others. Too much control will prevent us from seeing possibilities beyond our immediate desires. Too much order can stifle the restlessness of a truly open mind.

For the sake of democracy, we need to forge a new social compact—not a starry-eyed declaration of cyber-independence, but a realistic compromise between personal liberty and communal obligation. Government has a role to play, but increasingly individuals will have to balance their new power with new responsibilities to society at large. We must resist the urge to use technology to disengage from reality and its problems, using it instead to strengthen local communities and public discourse. Democracy may flourish in the era of individual control, but only if each of us makes the requisite sacrifices.

It won't be a primrose path. But with a little luck and a lot of work, we should be able to avoid the worst parts of this enigmatic and momentous development, even as we revel in the rest of it.

Afterword

If Internet time is to real time as dog years are to human years, then it has been nearly a decade since I finished writing the hardcover edition of this book in February 1999. Given the rapid pace of Internet developments over the past year, it certainly feels that way. In just the last dozen weeks or so, Microsoft was declared a monopoly by a U.S. judge, Y2K hysteria came and went, and media giant Time Warner agreed to be bought by America Online. Amid these developments, more subtle changes have also taken place indicating that we are living in a revolutionary time—one that has much to do with shifting power and one whose outcome is far from known.

Every week brings new signs that individuals are using the Internet to do things they could not do before. In the U.S., voters are perusing the web sites of presidential candidates or creating their own satires of those sites; they are making instant campaign contributions by credit card, and—in a few states like Arizona and Alaska—getting an opportunity to vote online in primaries. Consumers are using the Internet to compare computer prices instantly at hundreds of stores, to name the amount they'll pay for airline tickets, even to buy a car. Patients are confronting doctors with reams of medical information that they find online—or are bypassing their physicians altogether to obtain prescriptions and pharmaceuticals via the web. Outside the U.S., in places like China, millions of people are going online to get their first taste of freely available information and communication. New sites for specific ethnic and cultural groups are also linking people worldwide—for example, connecting Spanish speakers in Madrid, Mexico City, and Manhattan.

In each of these cases, individuals are being empowered—as citizens, as consumers—and there are plenty of entrenched elites who aren't happy about it. Car salesmen have protested auto manufacturers that allow purchasers to circumvent them (and their price markups). One shopping mall fearful of lost revenues over the winter holiday season launched an outright campaign against cyber-shopping. The Federal Election Commission has suggested that a homegrown web site favoring one or another political candidate should count as a campaign contribution—including the cost of the computer used to create the site. And Chinese authorities have announced an array of new laws meant to prevent the free flow of information and commerce —a "frontal assault on the Internet by a government that is afraid of losing control," in the words of the *New York Times*.

For the most part, this resistance to individual empowerment is wholly unwarranted, representing nothing more than fear, greed, or a desire to maintain power. Yet recent developments also show that there are some legitimate concerns about what the Internet makes possible. The Columbine High School students who shot their classmates to death last spring reportedly spent a lot of time visiting web sites filled with hateful, violent messages. Certainly they could have found those messages elsewhere, but the Internet made it easier for them to get access to those views and to others who espouse them. This doesn't mean that we should restrict what is available online, only that we should be ready to face the sometimes unpleasant consequences of life in a fully networked, information-rich world.

The dangers are not always grave. The convenience of hi-tech brings routine annoyances such as junk email, network failures, and the ability of the boss to get you anywhere on your cell phone. Nor are the problems always immediately apparent. Well-meaning citizens may use their new power in ways that can backfire in the long term. Federal and state law enforcement officials have recognized as much in trying to place some limits on sales of medicine online, an activity which will seem harmless only until we hear of people overdosing on Prozac—or Viagra. Similarly, the mall that protested Internet shopping may have been on to something, since far-away dot-com purchases could seriously erode the revenues of local stores, endangering jobs and economic well-being in areas that are not hubs of the so-called new economy.

Our response to these challenges, I have argued, must be to strike the right balance—between individual autonomy and connection to real communities, self-indulgence and commitment to others, traditional regulation and new means of promoting the public interest. This plea for moderation may seem almost radical given the anything-goes tenor of our society today. But the desire for absolute control may itself be an anomaly, a uniquely American quirk passed down to us by revolutionary forebears who understandably coveted autonomy. I say this in part because of a recent experience during a trip to Asia, where I had the opportunity to share some of the themes of this book with Japanese leaders from the public and private sectors. To all of them, balance and restraint were perfectly obvious responses to personal empowerment. Let us hope that the global reach of the Internet will allow Americans to borrow such wisdom—and others to borrow ours—so that we can together steer this revolution of technology and control toward fruitful outcomes.

Andrew L. Shapiro
New York City
March 2000

Notes

Introduction

1. Technorealism, a term coined by author David Shenk and myself, is a critical perspective on technology that is meant to go beyond the simple dualism of cyberutopianism and neo-Luddism. In early 1998, Shenk and I asked a handful of technology writers to join us in drawing up a set of principles that might begin to define technorealism (see www.technorealism.org). This book is, in a sense, an attempt to continue that effort.

Chapter 1: "We Have Revolution Now"

1. Miguel de Cervantes, *Don Quixote,* trans. John G. Lockhart (Edinburgh: A. Constable, 1822).
2. Quoted in Reuters, "The Soviet Crisis: Bush Monitors 'Serious Situation' as U.S. Scrambles for Information," *New York Times,* August 19, 1991.
3. Elizabeth L. Eisenstein, *The Printing Revolution in Early Modern Europe* (New York: Cambridge University Press, 1983), 150 (quoting German historian Johann Sleidan).
4. Stewart Brand, *The Media Lab: Inventing the Future at MIT* (New York: Viking, 1987), 213.
5. See Alvin Toffler, *Powershift: Knowledge, Wealth, and Violence in the 21st Century* (New York: Bantam Books, 1990), 356–358.
6. The Radio B92 broadcasts are still online at www.xs4all.nl/~opennet/audio.html
7. Chris Hedges, "Serbs' Answer to Oppression: Their Web Site," *New York Times,* December 8, 1996, A1. See also Laura Silber and Veran Matic, "Radio Free Yugoslavia," *New York Review of Books,* November 6, 1997, 67.
8. By that time, Voice of America and the BBC were already rebroadcasting their feed from the Net. "Focus on Serbia," *Atlanta Journal and Constitution,* December 17, 1996, A16.
9. Quoted in Hedges, "Serbs' Answer," A1. The Zapatistas in Mexico are another political group who have used the Internet effectively to gain attention for their cause.
10. Jasminka Udovicki and James Ridgeway, "The Revolution Might Be Televised," *Village Voice,* December 24, 1996, 36.
11. David Bennahum, "The Internet Revolution," *Wired,* April 1997, 168.
12. Sometimes a person's actions may simply reflect the will of an institution, such as a company for which he works or a government he serves. At the same time, institutional behavior always reflects the conduct of particular individuals, usually powerful elites.
13. The idea that technology will inevitably empower individuals can be found in writings such as Walter B. Wriston, *The Twilight of Sovereignty: How the Informa-*

tion Revolution is Transforming Our World (New York: Charles Scribner's Sons, 1992); "Cyberspace and the American Dream: A Magna Carta for the Knowledge Age," Release 1.2, August 22, 1994, at www.pff.org/position.html; and James Dale Davidson and Lord William Rees-Mogg, *The Sovereign Individual: How To Survive and Thrive During the Collapse of the Welfare State* (New York: Simon and Schuster, 1997).

14. The term "control revolution," though new to the context of the Internet and other new communications tools, has been used before in the realm of information and technology. In a 1986 book, James R. Beniger used the term to describe the transition from the industrial age to the information age that began more than a century ago. Actually, Beniger referred to four control revolutions, the first three being: (1) "molecular programming" indicating the origin of life, (2) learning by imitation (or culture), and (3) development of bureaucratic organizations. James R. Beniger, *The Control Revolution: Technological and Economic Origins of the Information Society* (Cambridge, Mass.: Harvard University Press, 1986), 62–63. The fourth was the advent of microprocessing and computing systems. Beniger argued that the complexity that accompanied the industrialization of the mid-nineteenth century spurred a "crisis of control," which could only be solved by the development of new control technologies in the areas of production, distribution, and consumption. Bureaucracy, he said, was the leading solution to that crisis of control. Along with new manufacturing processes, transportation methods, and communications technologies, it paved the way toward our information society of today. Beniger's focus was novel and the detailed historical explications in his book were impressive. But his analysis was fairly apolitical. He discussed control mechanisms with little or no attention to who decided how they were used. Were transportation and communications technologies in the hands of governments, corporations, or individuals? Were they deployed in ways that were democratic or authoritarian? Even putting aside the different time frame of his study, then, Beniger's control revolution is quite distinct from the one I am discussing here, which is specifically a shift in *who* controls information, experience, and resources, not just *how* those aspects of life are controlled.

15. Before Marshall McLuhan opined that the medium is the message, his mentor, Harold Innis, argued that some media, like books, are durable and allow us to disseminate knowledge over time, while others, like radio, are ephemeral and are thus better suited to the spread of information over a large space. Harold Innis, *The Bias of Communication* (Buffalo: University of Toronto Press, 1951), 33. The Internet is the first technology that allows us to do both. Expression in the form of a book cannot easily conquer the limits of space because the form is heavy and expensive to produce and distribute. But online, a book-length file can be sent anywhere immediately and sending a thousand of those files costs the same as sending one. Expression that occurs on radio or television cannot easily conquer the limits of time because it is presented at a fixed hour and is evanescent. Tape recorders and VCRs represent our desire to make it otherwise, but are really only stopgap measures. In cyberspace, audio and video can be called up whenever the user wants and the digital form of storage is ideally suited for long-term preservation.

16. John Tierney, "Our Oldest Computer, Upgraded," *New York Times,* September 28, 1997, 46.

17. Jeremy Rifkin, "The Biotech Century: Human Life as Intellectual Property," *Nation*, April 13, 1998, 11, 12.

CHAPTER 2: THE POLITICS OF CODE

1. In a debate over a failed bill to require handgun manufacturers to install child-proof "trigger locks," Senator Barbara Boxer noted that 440 children die each year as a result of accidental shootings and eight times that number are injured. Edward Chen, "Senate Rejects Bid to Require Gun Safety Locks," *Los Angeles Times*, July 22, 1998, A3.

2. McLuhan, in classic understatement, put the point this way: "Our conventional response to all media, namely that it is how they are used that counts, is the numb stance of the technological idiot." Marshall McLuhan, *Understanding Media: The Extensions of Man* (New York: McGraw-Hill, 1964), 18. For more nuanced discussions of whether technologies are political, see Langdon Winner, *The Whale and the Reactor: A Search for Limits in an Age of High Technology* (Chicago: University of Chicago Press, 1986); Langdon Winner, *Autonomous Technology: Technics-Out-of-Control as a Theme in Political Thought* (Cambridge, Mass.: MIT Press, 1977); and Ithiel de Sola Pool, *Technologies of Freedom* (Cambridge, Mass.: Belknap Press, 1983).

3. My use of "the Net" will be inclusive of today's Internet, though I will use the unabbreviated term "Internet" when I want to refer specifically to that medium as it exists now.

4. See, for example, Report of Special Rapporteur Abid Hussain to the United Nations Commission on Human Rights, "Promotion and Protection of the Right to Freedom of Opinion and Expression," January 28, 1998, section III(C) (stating that "new technologies and, in particular, the Internet are inherently democratic").

5. See, for example, Lawrence Lessig, "The Law of the Horse: What Cyberlaw Might Teach," available, along with other articles dealing with the importance of code, at cyber.harvard.edu/lessig.html. MIT professor William J. Mitchell says in *City of Bits: Space, Place, and the Infobahn* (Cambridge, Mass.: MIT Press, 1995), 112: "Control of code is power. . . . Arcane text in highly formalized language, typically accessible to only a few privileged high-priests . . . is the medium in which intentions and designs are realized, and it is becoming a crucial focus of political contest." An even earlier expression of this idea was Mitchell Kapor's claim that the architecture of the Internet is political. See Mitchell Kapor, "Where Is the Digital Highway Really Heading? The Case for a Jeffersonian Information Policy," *Wired*, July/August 1993, 53.

6. The existence of conference calls and party lines does not change the predominantly one-to-one nature of telephony.

7. Pew Research Center for the People and the Press, "The Internet News Audience Goes Ordinary," January 14, 1999, at www.people-press.org/tech98sum.htm.

8. The examples in this paragraph are from Steven S. Miller, *Civilizing Cyberspace: Policy, Power, and the Information Superhighway* (Reading, Mass.: Addison-Wesley, 1996), 30; Pool, *Technologies of Freedom*, 27; Winner, *Whale and the Reactor*, 12.

9. In April 1998, one of the world's most distinguished research journals, *Science*,

published an article by two leading Internet researchers concluding that there were major disparities in Internet use between black and white Americans. The finding was deemed significant enough to merit coverage on the front page of the *New York Times*. Amy Harmon, "Racial Divide Found on Information Highway," *New York Times*, April 17, 1988, A1. But it was based on data collected during the winter of 1996/97. In the intervening year or so, the number of Internet users had doubled from about 30 million to almost 60 million, according to another study of Internet demographics, and the racial breakdown among those 60 million users directly paralleled that of the general U.S. population. See David Birdsell, Douglas Muzzio, David Krane, and Amy Cottreau, "Web Users Are Looking More Like America," *The Public Perspective*, April/May 1998, 33–35. The *Science* study was not wrong. It was just out of date by the time it was published.

10. White House Press Release, July 1, 1997 (Remarks by the President in Announcement of Electronic Commerce Initiative); Executive Summary, *A Framework for Global Electronic Commerce*, July 1, 1997.

11. White House Press Release, July 1. A November 1998 update from the Clinton administration reiterated the message. See www.doc.gov/ecommerce/review.htm.

12. The conventional wisdom is corrected in Katie Hafner and Matthew Lyons, *Where Wizards Stay Up Late: The Origins of the Internet* (New York: Simon and Schuster, 1996), 10, 77.

13. Alexis de Tocqueville, *Democracy in America,* ed. J. P. Mayer, trans. George Lawrence (New York: HarperPerennial, 1988), 508. Some of the more celebrated contemporary accounts of American individualism include Robert N. Bellah, Richard Madsen, William M. Sullivan, Ann Swidler, and Steven M. Tipton, *Habits of the Heart: Individualism and Commitment in American Life* (Berkeley: University of California Press, 1985), and David Riesman, *The Lonely Crowd: A Study of the Changing American Character* (New Haven, Conn.: Yale University Press, 1950).

14. Statistics in this paragraph appear in Joseph S. Nye Jr., "Introduction: The Decline of Confidence in Government," in *Why People Don't Trust Government,* eds. Joseph S. Nye Jr., Philip D. Zelikow, and David C. King (Cambridge, Mass.: Harvard University Press, 1997), 1–2.

15. Ibid.

16. Charles Handy, *The Hungry Spirit: Beyond Capitalism: A Quest for Purpose in the Modern World* (New York: Broadway Books, 1998), 67; see also Davidson and Rees-Mogg, *The Sovereign Individual*.

17. Isaiah Berlin, "Two Concepts of Liberty," in *Four Essays on Liberty* (New York: Oxford University Press, 1969), 131.

18. Stanley Renshon, *Psychological Needs and Political Behavior* (New York: Free Press, 1974), 6.

19. Quoted in Renshon, *Psychological Needs and Political Behavior,* 52.

20. Stanley A. Renshon, "The Need for Personal Control in Political Life: Origins, Dynamics, and Implications," in *Choice and Perceived Control,* eds. Lawrence C. Perlmuter and Richard A. Monty (Hillsdale, N.J.: Lawrence Erlbaum Associates, 1979), 41, 45. People are "happier, healthier, more active, solve problems better, and feel less stress when they are given choice and control." Martin E. P. Seligman and Suzanne M. Miller, "The Psychology of Power: Concluding Comments," in *Choice and Perceived Control,* 347, 368.

21. Renshon, "The Need for Personal Control," 46–47.
22. See David Brin, *The Transparent Society: Will Technology Force Us to Choose Between Privacy and Freedom?* (Reading, Mass.: Addison-Wesley, 1998), 155.
23. Herbert M. Lefcourt, *Locus of Control: Current Trends in Theory and Research* (Hillsdale, N.J.: Erlbaum Associates, 1982), 2.
24. Diane B. Arnkoff and Michael J. Mahoney, "Perceived Control and Psychopathology," in *Choice and Perceived Control,* 155, 159.
25. Indeed, our continuing quest for control should remind us that we are never that far from the archetypal child yearning to make his own way: "At each stage in the unfolding discovery of his world and the problems that it poses, the child is intent on shaping the world to his own aggrandizement. He has to keep the feeling that he has absolute power and control, and in order to do that he has to cultivate independence of some kind, the conviction that he is shaping his own life." Ernest Becker, *The Denial of Death* (New York: The Free Press, 1973), 37. For a popular treatment of issues of control in everyday life, see Judith Viorst, *Imperfect Control: Our Lifelong Struggle with Power and Surrender* (New York: Simon and Schuster, 1998).

Chapter 3: Gaining Control

1. Bill Gates, *The Road Ahead,* rev. ed. (New York: Penguin, 1996), 1–2.
2. Quoted in Steven Levy, *Hackers: Heroes of the Computer Revolution* (Garden City, N.Y.: Anchor Press/Doubleday, 1984), 284.
3. "Many hobbyists used the kind of control they felt able to achieve with their home computers to relieve a sense that they had lost control at work and in political life." Sherry Turkle, *Life on the Screen: Identity in the Age of the Internet* (New York: Simon and Schuster, 1995), 32.
4. Quoted in Molly O'Neill, "The Lure and Addiction of Life On Line," *New York Times,* March 8, 1995, C1, C6.
5. Turkle cites the case of a middle-aged lawyer who is passionate about playing games on his computer because, he says, "with these games I'm in complete control. It's a nice contrast with the rest of my life." Turkle, *Life on the Screen,* 67.
6. Renshon, *Psychological Needs,* 49–50.
7. Steven Johnson, *Interface Culture: How New Technology Transforms the Way We Create and Communicate* (San Francisco: HarperEdge, 1997), 21.
8. The ad can be found in the June 1995 issue of *Wired.*
9. John Perry Barlow, Declaration of the Independence Cyberspace, February 9, 1996 (at www.eff.org/pub/Publications/John_Perry_Barlow/barlow_0296.declaration).
10. See, for example, Howard Rheingold, *The Virtual Community: Homesteading on the Electronic Frontier* (Reading, Mass.: Addison-Wesley, 1993).
11. David R. Johnson and David Post, "Law and Borders: The Rise of Law in Cyberspace," *Stanford Law Review* 48 (1996): 1367, 1378.
12. "We have decided to call the entire field of control and communications theory . . . by the name Cybernetics," Norbert Wiener wrote in 1948. Norbert Wiener, *Cybernetics: or Control and Communication in the Animal and the Machine* (Cambridge, Mass.: MIT Press, 1961), 11. Wiener himself may have been unaware that 150 years earlier a French physicist also used the term cybernetics to refer to a

branch of political science which he described as the science of governance. See Kevin Kelly, *Out of Control: The Rise of Neo-Biological Civilization* (Reading, Mass.: Addison-Wesley, 1994), 120.

The term cyberspace was first used by William Gibson, in his science-fiction story "Burning Chrome," *Omni,* July 1982, 72. It was made popular in his novel *Neuromancer* (New York: Ace Books, 1984).

13. See Donald A. Norman, *The Invisible Computer: Why Good Products Can Fail, the Personal Computer is So Complex, and Information Appliances Are the Solution* (Cambridge, Mass.: MIT Press, 1998).

14. To a degree, television has played this role over the last few decades, influencing our personal and social life, our understanding of issues, and our political system. In fact, a small precursor to the control made possible by the Net can be seen in the changes wrought by the television remote control. That little clicker meant that rather than watching TV passively, we could shuttle between programs, mute the volume, and avoid commercials (often to the consternation of broadcasters and advertisers). It's a shift that is being recapitulated on a much more dynamic scale in our use of the Net. Like the TV remote, the networked computer gives us more power, only now the experience we can control is not just television, but almost the full scale of our human interactions.

CHAPTER 4: LIEBLING'S REVENGE

1. Philip Elmer-DeWitt, "On a Screen Near You: Cyberporn," *Time,* July 3, 1995, 38.

2. Elmer-DeWitt was not the only one duped by Rimm's research. The *Georgetown Law Journal* agreed to publish the study before *Time* did, despite the fact that it was not peer reviewed. See Marty Rimm, "Marketing Pornography on the Information Superhighway: A Survey of 917,410 Images, Descriptions, Short Stories, and Animations Downloaded 8.5 Million Times by Consumers in Over 2000 Cities in Forty Countries, Provinces, and Territories," *Georgetown Law Journal* 83 (1995): 1849.

3. These comments are from the far-ranging discussion that occurred on the Well. See www.hotwired.com/special/pornscare/well.

4. Brock Meeks, "How *Time* Failed," *Hotwired, at* http://www.hotwired.com/special/pornscare/brock.html (originally published in Meeks's CyberWire Dispatch, July 7, 1995).

5. See Philip Elmer-DeWitt, message posted to the Well, July 21, 1997, at 02:51. (My thanks to Mike Godwin for pointing out this source to me.)

6. Ibid.

7. Philip Elmer-DeWitt, "Fire Storm on the Computer Nets: A New Study of Cyberporn, Reported in a *Time* Cover Story, Sparks Controversy," *Time,* July 24, 1995, 57.

8. The controversy was ultimately covered in the *New York Times* and the *Washington Post,* among other national news outlets. Peter H. Lewis, "The Internet Battles a Much-Disputed Study on Selling Pornography on Line," *New York Times,* July 17, 1995, D5; Howard Kurtz, "A Flaming Outrage; A 'Cyberporn' Critic Gets a Harsh Lesson in '90s Netiquette," *Washington Post,* July 16, 1995, C1.

9. This maelstrom also showed online publishing at its best. HotWired, then a leading web site, dedicated a special area to the controversy with the equivalent of

hundreds of pages of text. There were links to other related sites and, with essentially no space limits, HotWired could present unedited versions of the exchanges that had transpired on the Well. In a rather unusual move, one media outlet was devoting substantial resources to discrediting the reporting of another, and it was doing so by relying on feedback from everyday citizens as well as from journalists. (These materials are still up at www.hotwired.com/special/pornscare.)

Episodes like this are emblematic of a new form of journalism on the web, which might be called "layered" coverage. Stories start brief and general, but readers are offered the opportunity to dig down to deeper layers that are increasingly more detailed and expansive. Layered coverage is about more than just the endless space that online publishing provides. It's about the way that interactivity changes the experience of news for each individual. Newspapers and TV news may try to do the same. On the front page or at the top of the hour, they may give a brief teaser for a story presented in full later. But they cannot offer readers the range and versatility—and ultimately the user empowerment—that online journalism can.

Layered coverage allows for innovative types of public-service journalism. In February 1997, NBC's *Dateline* ran a program on road safety, highlighting the hazards presented on certain dangerous roads throughout America. Viewers were told that they could log on to NBC's online counterpart, MSNBC, to find out about dangerous roads in their own areas. All they needed to do was enter their zip code and, instantly, they could receive federal data on the number of fatalities that had occurred on roads in their vicinity. Tens of thousands of viewers did so. *Money* magazine showed another way to use the web to make the news more relevant to each reader. When the "flat tax" was being heralded by presidential candidate Steve Forbes, *Money*'s online site allowed users to enter personal income data and find out how a flat tax might affect them. See John V. Pavlik, "The Future of Online Journalism: Bonanza or Black Hole?" *Columbia Journalism Review*, July/August 1997, 30–31.

10. Quoted in Lewis, "The Internet Battles." Godwin's own account of the *Time* cyberporn incident can be found in chapter 9 of his book *Cyber Rights: Defending Free Speech in the Digital Age* (New York: Times Books, 1998).

11. Quoted in Joshua Quittner, "Man Bites Web," *Time*, October 21, 1996, 70.

12. Howard Kurtz, "A Flaming Outrage."

13. A. J. Liebling, *The Press* (New York: Ballantine Books, 1975), 32.

14. See Ben H. Bagdikian, *Media Monopoly* (Boston: Beacon Press, 1997); Robert McChesney and Edward S. Herman, *The Global Media: The New Missionaries of Corporate Capitalism* (Washington, D.C.: Cassell, 1997).

15. These critiques, though often associated with commentators on the left, actually have come from observers across the spectrum. In 1969, for example, then Vice President Spiro Agnew announced that "the American people should be made aware of the trend toward the monopolization of the great public information vehicles and the concentration of more and more power in fewer and fewer hands." *New York Times*, November 21, 1969, 1.

16. Quoted in J. D. Lasica, "Goodbye, Gutenberg?" American Journalism Review/NewsLink, July 7–13, 1998, at www.newslink.org/ajrjd4.html.

17. See Panos Institute, "The Internet and the South: Superhighway or Dirt-Track?"

October 1995, at www.oneworld.org/panos/briefing/internet.htm. This study also maintains that 49 countries have fewer than one phone line per hundred people and that 80 percent of the world population lacks basic telecommunications, such as phone service. See also William Wresch, *Disconnected: Haves and Have-nots in the Information Age* (New Brunswick, N.J.: Rutgers University Press, 1996).

18. In the U.S. in 1996, the most recent year for which data are available, 98 percent of households had televisions and 94 percent had telephone service; 82 percent of homes with television had VCRs, while 65 percent had cable television access. U.S. Census Bureau, *Statistical Abstract of the United States: 1998*, 573 (1998). Telephone penetration rates in the rest of the developed world are also around the 94 percent mark. International Telecommunications Union, *World Telecommunication Development Report 1998: Universal Access* (March 1998), 15.

19. See www.liszt.com.

20. See www.dfn.org.

21. See www.bolt.com.

22. The new MP3 format of digital music storage, in fact, has begun to cause consternation in the recording industry. See Jon Pareles, "Trying to Get in Tune With the Digital Age," *New York Times*, February 1, 1999, C1.

23. Already, individuals are using the web to distribute video—some on a twenty-four-hour-a-day basis. Jennicam (www.jennicam.org), for example, is a site that shows a live stream of images from the Washington, D.C., home of a young woman named Jennifer Ringley. You can watch Jenni get dressed in the morning, talk to her friends on the phone, sleep, and so on. Ringley, a pioneer in the webcam arena with thousands of fans, has apparently inspired many others to put their lives online. See Rick Marin, "And Now, the Human Show," *Newsweek*, June 1, 1998, 64.

CHAPTER 5: MASTERS OF OUR OWN DOMAINS

1. Michel Marriot, "Happy Birthday, Your Team Won, Your Stock Crashed," *New York Times*, September 17, 1998, G1.

2. David Shenk, *Data Smog: Surviving the Information Glut* (New York: Harper-Collins, 1997), 35–50.

3. As of late 1998, nearly one in four experienced Internet users receives customized news by email. And more than 40 percent of experienced users use the Net to get personalized information such as stock quotes. Pew Research Center for the People and the Press, "The Internet News Audience Goes Ordinary."

4. Nicholas Negroponte, *Being Digital* (New York: Knopf, 1995), 153.

5. Anthony Smith, *Goodbye, Gutenberg: The Newspaper Revolution of the 1980s*, quoted in Brand, *The Media Lab*, 38.

6. Adam Clayton Powell III, "*Jerusalem Post* Goes 'Paper View,'" at www.freedomforum.org/technology/1997/12/22jerupost.asp.

7. Anne Eisenberg, "If the Shoe Fits, Click It," *New York Times*, August 13, 1998, D1.

8. Jessie Scanlon, "Have It My Way," *Wired*, April 1997, 70.

9. Daniel Pink, "Free Agent Nation," *Fast Company*, December/January 1998, 131, 132.

10. Federal News Service, "Remarks by President Clinton and Vice President Al Gore Regarding the Administration Report on Electronic Commerce," July 1, 1997.

11. When an online research or entertainment tool is set up without a chat space, for

example, users will often demand one—or figure out some other way to communicate with one another. They want to establish social ties around the activity that brought them to the same site in the first place.

12. Quoted in Harvey Blume, "On the Net," *New York Times*, June 30, 1997, D6. See also On the Same Page, at amug.org/~a2o3/index.html.

13. Quoted in William Glaberson, "A Guerilla War on the Internet," *New York Times*, April 8, 1997, B1.

14. Quoted in A. Lin Neumann, "The Resistance Network," *Wired*, January 1996, 108, 110; see also Michael Marriot, "Amplifying Voices for Human Rights," *New York Times*, February 26, 1998, G14.

15. Examples include Lotus Marketplace, Lexis P-Trak, and America Online's mid-1997 plan to sell subscribers' phone numbers. See Karen Kaplan, "AOL Drops Plans to Sell Members' Phone Numbers," *Los Angeles Times*, July 25, 1997, A1; Laura J. Gurak, *Persuasion and Privacy in Cyberspace: The Online Protests over Lotus Marketplace and the Clipper Chip* (New Haven: Yale University Press, 1997).

16. Graeme Browning, *Electronic Democracy: Using the Internet to Affect American Politics* (Wilton, Conn.: Pemberton Press Books/Online Inc., 1996), 53.

17. Constance Hale, "How Do You Say Computer in Hawaiian?" *Wired*, August 1995, 90. Latin language scholars worldwide have similarly used the Net to communicate in their favorite idiom. "There are so few people scattered around the world who have any interest in actually speaking or writing Latin," says Jeffrey Wills, a classics professor at University of Wisconsin in Madison. "But you could establish critical mass by getting them together on the Internet." Steve Coates, "Et Tu, Cybernetica Machina User?" *New York Times*, October 28, 1996, D4.

18. Sherry Turkle, *Life on the Screen*, 255–269.

Chapter 6: The Decline of Middlemen

1. Statement by Chairman Arthur Levitt Concerning On-line Trading, January 27, 1999, at www.sec.gov/news/press/99-9.txt.

2. Sana Siwolop, "Online Investors Chase Every Blip," *New York Times*, October 1, 1998.

3. Mark Gimein, "Around the Globe, Net Stock Mania," *Industry Standard*, December 21, 1998; Ken Maney, "Birth of a New Order," *USA Today*, December 31, 1998, B1.

4. Recognizing the masculine nature of the word middleman, I will use it to refer to intermediaries regardless of gender because it is a standard term in the field.

5. Some full-time day traders actually get access to a stock exchange's internal trading system, such as Nasdaq's Small Order Exchange System. See Siwolop, "Online Investors Chase Every Blip."

6. See www.virtualvineyards.com.

7. See www.pricescan.com.

8. Microsoft's Expedia site, an online travel agent, blocks agent software that tries to compare prices. See Scott Woolley, "Bricks and Mortar Fight Back," *Forbes*, September 21, 1998, 256, 257.

9. Quoted in Jon Christensen, "A New On-line Publisher Promotes Nonfiction on a Pay-Per-Read Basis," *New York Times*, September 22, 1997, D9.

10. Robert Dahl, *Democracy and Its Critics* (New Haven: Yale University Press, 1989), 18–20.

11. Buckminster Fuller, *No More Secondhand God, And Other Writings* (Carbondale, Ill.: Southern Illinois University Press, 1963), 12–14.

12. Thomas E. Cronin, *Direct Democracy: The Politics of Initiative, Referendum, and Recall* (Cambridge, Mass.: Harvard University Press, 1989), 220.

13. See, for example, www.vote.org.

14. See, for example, Davidson and Rees-Mogg, *The Sovereign Individual*, 35 ("representative democracy as it is now known will fade away, to be replaced by the new democracy of choice in the cybermarketplace"); see also John Naisbitt, *Megatrends* (New York: Warner Books, 1982).

15. See "Life at the Democratic Roots," *Economist*, December 21, 1996, S7.

CHAPTER 7: AN ANXIOUS STATE

1. Michel Foucault, *The History of Sexuality, Volume 1: An Introduction,* trans. Robert Hurley (New York: Vintage Books, 1978), 95.

2. These issues are discussed intelligently in recent publications including Michael R. Nelson, "Sovereignty in the Networked World," in *The Emerging Internet* (Queenstown, Md.: Aspen Institute, Institute for Information Studies, 1998), 1, and Jack L. Goldsmith, "Against Cyberanarchy," *University of Chicago Law Review* 65 (1998): 1199.

3. The Internet would hardly be the first technology to cause such a row. As Ithiel de Sola Pool observed, "Each new advance in the technology of communications disturbs a status quo. It meets resistance from those whose dominance it threatens . . . ." Pool, *Technologies of Freedom,* 7.

4. See Global Internet Liberty Campaign, "Regardless Of Frontiers: Protecting The Human Right to Freedom of Expression on the Global Internet," September 5, 1998, at www.gilc.org/speech/report/

5. See Human Rights Watch, "Silencing the Net: The Threat to Freedom of Expression On-line" (May 1996), at www.cwrl.utexas.edu/~monitors/1.1/ [hereafter HRW Report]; Canadian Committee to Protect Journalists, "Internet Censorship Report: The Challenges for Free Expression On-line" (April 1998), at www.ccpj.ca/publications/internet/index.html [hereafter CCPJ Report]. See also the GILC Alerts archived at www.gilc.org/alert.

6. See Sheila Tefft, "China Attempts to Have Its Net and Censor It Too," *Christian Science Monitor,* August 5, 1996, 1; HRW Report.

7. Geremie R. Barme and Sang Ye, "The Great Firewall of China," *Wired,* June 1997, 138, 147.

8. CCPJ Report.

9. Joe McDonald, "Chinese Man Sentenced for Email," Associated Press, January 20, 1999.

10. CCPJ Report; see also www.sba.gov.sg/internet.htm.

11. CCPJ Report.

12. See Matthew McAllester, "For Many, the Internet is Access to Democracy," *Newsday,* November 12, 1997, C12.

13. HRW Report.

14. Nghiem Xuan Tinh, quoted in Jeremy Grant, "Vietnamese Move to Bring the Internet Under Control May Backfire," *Financial Times* (London), September 19, 1995, 6.

15. Reuters, "Saudis to Get Censored Version of Internet," November 5, 1997.

16. Agence France Presse, "Kuwait to Boost Internet Service but Plans Censorship," April 11, 1996.

17. HRW report.

18. Ulrich Sieber, quoted in Edmund L. Andrews, "Germany Charges Compuserve Manager," *New York Times*, April 17, 1997, D19.

19. Perhaps if it had been put on notice, Compuserve could have tried to completely restructure its service to offer the kind of sanitized Internet access that China and other nations attempt to offer. Yet, in addition to being authoritarian and incredibly costly, such a plan would not really work unless the Bavarian government required every service provider to do the same—thus effectively turning Bavaria into another China.

20. David Hudson, "Germany's Internet Angst," Wired News, May 28, 1998, at www.wired.com/news/news/email/other/politics/story/12884.html; "CompuServe Ruling Awaited," *National Law Journal*, June 22, 1998, A14.

21. Marquardt, disclaiming any support for *Radikal* and its supposedly violent tactics, maintained that she was linking to the site merely as an act of free expression. "She wants to have a discussion about censorship, but to discuss censorship you have to be able to show people the kinds of newspapers that might be censored," her attorney said. Volker Ratzmann, quoted in Edmund L. Andrews, "German Judge Dismisses Criminal Charge Over Internet Link," *New York Times*, July 1, 1997, D7.

22. Levi Rizetnikof, "New Scapegoat: The Internet," *Wired*, July 1995, 33.

23. See *Reno v. American Civil Liberties Union*, 521 U.S. 844 (1997).

24. *Ginsberg v. New York*, 390 U.S. 629 (1968).

25. *FCC v. Pacifica*, 438 U.S. 726 (1978).

26. One wonders whether such an extreme governmental response might have been motivated in part by the radical statements noted earlier that nation-states have no jurisdiction over online interactions.

27. Congress, for example, held no hearings regarding the Internet before passing the CDA. See *Reno*, 521 U.S. at 889 n.23.

28. This type of code regulation fits the model of power described by the French social theorist Michel Foucault. Foucault warns that state power in modern society is increasingly hidden in the structure of everyday life. For example, in considering state use of surveillance technology, Foucault says that what makes it so powerful is the way it changes our expectations of privacy and ultimately our behavior—all without traditional coercion. See Michel Foucault, *Discipline and Punish: The Birth of the Prison*, ed. and trans. Alan Sheridan (New York: Pantheon Books, 1979); see also James Boyle, "Foucault in Cyberspace, Surveillance, Sovereignty, and Hardwired Censors," *University of Cincinnati Law Review* 66 (1997): 177.

29. History shows us similar examples of increasingly sophisticated resistance. When printing emerged, the British crown responded between the sixteenth and eighteenth centuries with a variety of constraints, one more cleverly fine-tuned than

the next: from outright censorship and the grant of monopoly printing rights to licensing requirements, taxation, and libel actions by the state against "seditious" publications. Over time, advocates of freedom succeeded in exposing all of these as unjust instruments of repression. But the struggles were not easy. Nor will they be easy today and in the future.

30. Peter Wayner, "Giving Away Secrets," *New York Times*, July 29, 1997, A19.

31. For more on public-key cryptography, see Whitfield Diffie and Susan Landau, *Privacy on the Line: The Politics of Wiretapping and Encryption* (Cambridge, Mass.: MIT Press, 1998), 36.

32. "Secrecy is a form of power," law professor Michael Froomkin writes. "The ability to protect a secret, to preserve one's privacy, is a form of power. The ability to penetrate secrets, to learn them, to use them, is also a form of power." Froomkin, "The Metaphor is the Key: Cryptography, the Clipper Chip, and the Constitution," *University of Pennsylvania Law Review* 143 (1995): 709, 712. In this sense, the battle over encryption policy is a power struggle over who controls information: individuals or governments.

33. Quoted in Brin, *The Transparent Society*, 273.

34. Louis J. Freeh, "The Impact of Encryption on Public Safety," Statement Before the Permanent Select Committee on Intelligence, United States House of Representatives, September 9, 1997.

35. Diffie and Landau, *Privacy on the Line*, vii.

36. A similarly blunt effort was made by the federal government with the Communications Assistance Law Enforcement Act (CALEA), which requires U.S. phone companies to change the technical architecture of their networks so that law enforcement officials can easily impose wiretaps on digital phone lines. See, for example, Jim Warren, "Surveillance-on-Demand," *Wired*, February 1996, 72.

37. A federal district court has ruled that the federal government's encryption export regulations constitute an unconstitutional prior restraint on free speech. *Bernstein v. United States Department of State*, 974 F. Supp. 1288 (N.D. Cal. 1997).

38. Louis J. Freeh, Statement Before the U.S. Senate Committee on Commerce, Science, and Transportation Regarding the Impact of Encryption of Law Enforcement and Public Safety, March 19, 1997, 5.

39. See, for example, Edmund L. Andrews, "U.S. Restrictions on Exports Aid German Software Maker," *New York Times*, April 7, 1997, D1.

40. See Committee to Study National Cryptography Policy, Computer Science and Telecommunications Board, National Research Council, "Cryptography's Role in Securing the Information Society," May 30, 1996, at www2.nas.edu/cstb-web/2646.html (recommendation 2). The authors of the National Research Council report were given unprecedented access to classified information that purportedly would show why key escrow was necessary to protect national security. Yet these independent experts—among them many former government officials—disagreed with that conclusion. See also the web site of Americans for Computer Privacy, www.computerprivacy.org.

41. Congress is empowered "to promote the Progress of Science and useful Arts, by securing for limited Times to Authors and Inventors the exclusive Right to their respective Writings and Discoveries." U.S. Constitution, Art. I, sec. 8.

42. Fair use, the Supreme Court has held, is more likely to be found where an origi-

nal work is transformed in some way—for example, by the manner or context in which the work is used. See *Campbell v. Acuff-Rose*, 510 U.S. 569 (1994).

43. A large publisher concerned about the future of intellectual property, for example, speaks of "the anxiety all of us feel in the digital world." Charles R. Ellis, chairman of John Wiley and Sons, quoted in Doreen Carvajal, "An Electronic Sheriff to Battle Book Rustling," *New York Times*, September 22, 1997, D1.

44. The term "copyright maximalist" was coined by Professor Pam Samuelson. See Pam Samuelson, "The Copyright Grab," *Wired*, January 1996, 134, and "Protecting User Interfaces Through Copyright: The Debate," in *Proceedings of the ACM Conference on Computer-Human Interaction* (New York: Association for Computing Machinery, 1989), 97.

45. Among other things, the white paper proposed that most every digital transmission of information be considered a protected copy, even those temporary copies on a personal computer that are necessary, for example, to view material on the web. Denise Caruso, "A Tough Stance on Cyberspace Copyrights," *New York Times*, January 19, 1998, D3.

46. Denise Caruso, "Global Debate Over Treaties on Copyright," *New York Times*, December 16, 1996, D1. Copyright generally protects creative expression, not the raw information that goes into that expression. See *Feist Publications v. Rural Telephone Service Co.*, 499 U.S. 34 (1991). Unless limited substantially, database protection could alter the balance of intellectual property law and thus diminish much of the individual access to information that the control revolution makes possible. It would also, in the words of a joint statement by leading U.S. scientific organizations, "seriously undermine the ability of researchers and educators to access and use scientific data, and would have a deleterious impact on our nation's research capabilities." Statement of National Academy of Sciences, the National Academy of Engineering, and the Institute of Medicine, quoted in Pam Samuelson, "Confab Clips Copyright Cartel," *Wired*, March 1997.

47. Copyright protection now generally lasts for the life of the author plus seventy years (instead of fifty years). See Sonny Bono Copyright Term Extension Act, Pub. L. No. 105–298, 112 Stat. 2827 (1998) (codified in various sections of title 17 of the United States Code).

48. Andrew Ross Sorkin, "Digital 'Watermarks' Assert Internet Copyright," *New York Times*, June 30, 1997, D11; Carvajal, "An Electronic Sheriff."

49. At least one federal court has already suggested that clickwrap contracts are enforceable. See *Hotmail Corp. v. Van Money Pie Inc.*, 1998 U.S. Dist. Lexis 10729 (N.D. Ca. 1998); Martin H. Samson, "Click-Wrap Agreement Held Enforceable," *New York Law Journal*, June 30, 1998, 1. The likelihood that other courts will follow suit is based partly on *ProCD, Inc. v. Zeidenberg*, 86 F.3d 1447 (7th Cir. 1996), a case in which the U.S. Court of Appeals for the Seventh Circuit upheld a similar software "shrinkwrap license," overturning a district court's ruling that it was unenforceable because consumers didn't have an opportunity to negotiate its terms.

50. See Mark Stefik, "Trusted Systems," *Scientific American*, March 1997, 4.

51. The law contained some limited exemptions for libraries and encryption research. See Digital Millennium Copyright Act, 17 U.S.C. sec. 1201(d), (g) (1998).

52. See Julie E. Cohen, "A Right to Read Anonymously: A Closer Look at 'Copyright Management' in Cyberspace," *Connecticut Law Review*, 28 (1996): 981.

Chapter 8: Where Do You Want to Go Today?

1. The campaign was announced in late 1994, but didn't become ubiquitous until 1995. See PR Newswire, "Microsoft Journeys on Global Brand Campaign," November 10, 1994, available on Lexis/Nexis.

2. The Internet, says Gates, "allows people to go wherever they want to go. . . . The kind of control and flexibility that people have always wanted are finally delivered by these tools." Bill Gates, Public Lecture Sponsored by the 92nd Street Y, New York City, November 27, 1995 [hereafter November 1995 Public Lecture].

3. See Andrew McMains, "A Study in Contrasts," *Adweek*, February 1, 1999, 5; Todd Wallack, "Lotus Hopes Customers Get the Message," *Boston Herald*, January 19, 1999, 30.

4. See Kate Fitzgerald, "AT&T Dwells on the Future in New Ads," *Advertising Age*, April 26, 1993, 39; Kate Fitzgerald, "AT&T Ads Again Look at Future," *Advertising Age*, May 9, 1994, 8.

5. Some companies, to be sure, thrive off of the new user interactivity. Online vendors such as Amazon and Dell, for example, have seen revenues soar because of the ability of individuals to buy books and computers directly online. So have companies like GeoCities and Tripod, which host personal web pages built by individuals.

6. Jean-Jacques Rousseau, *Emile,* trans. Allan Bloom (New York: Basic Books, 1979), 120. For more on the ways in which corporations present consumers with dubious messages of individual empowerment, see Leslie Savan, *The Sponsored Life: Ads, TV, and American Culture* (Philadelphia: Temple University Press, 1994); Thomas Frank, *The Conquest of Cool: Business Culture, Counterculture, and the Rise of Hip Consumerism* (Chicago: University of Chicago Press, 1997).

7. Gates, November 1995 Public Lecture.

8. See "Microsoft's Trophies," *Time,* June 1, 1998, 36; Nathan Newman, "From Microsoft Word to Microsoft World: How Microsoft is Building a Global Monopoly," NetAction White Paper, 1997, 6–7, at www.netaction.org.

9. David Bank, "Microsoft Moves to Rule On-line Sales," *Wall Street Journal,* June 5, 1997, B1.

10. See Andrew L. Shapiro, "Hard Drive on Microsoft," *Nation,* December 8, 1997, 22, 24.

11. Bank, "Microsoft Moves."

12. See Complaint, *United States v. Microsoft Corp.*, July 15, 1994, D.D.C. (No. 94–1564). At the time, the relevant operating system could have been either MS-DOS or Windows, but for convenience sake I will refer in the text only to Windows.

13. Rockefeller pressured the railroad companies into a deal where that he would receive a rebate not only on every barrel of Standard Oil shipped, but on every barrel of a competitor's oil, as well. Ron Chernow, Rockefeller's biographer, calls this "an instrument of competitive cruelty unparalleled in industry." See Jack Beatty, "A Capital Life" (review of Ron Chernow's *Titan), New York Times Book Review,* May 17, 1998, 10.

14. The relevant portion of the consent decree, section IV(E)(i), is cited in *U.S. v. Microsoft Corp.*, 147 F.3d 935, 939 (D.C. Cir. 1998).

15. Steve Lohr and John Markoff, "How a Giant Software Maker Played the Game of Hardball," *New York Times*, October 8, 1998, A1.

16. See, for example, John Markoff, "Oracle Deal for Netscape's Network Computing Unit Is Seen," *New York Times*, May 19, 1997, D6; see also Laurence Zuckerman, "I.B.M. Tries to Outmaneuver Microsoft by Supporting Java," *New York Times*, September 1, 1997, D1.

17. Critics saw confirmation of this in the fact that Sun sued Microsoft in late 1997 for trying to create a proprietary version of Java, in violation of contract agreements. As of this writing, Sun had succeeded in obtaining a preliminary injunction in its suit. See *Sun Microsystems, Inc. v. Microsoft Corp.*, 21 F. Supp. 2d 1109 (N.D. Cal. 1998).

18. Quoted in David Chun, "Required to Buy Windows," at www.essential.org/antitrust/ms/jun3survey.html.

19. The operating system Linux generally has to be installed after the purchase of a PC, though this may soon change.

20. See Brian Arthur, "Increasing Returns and the New World of Business," *Harvard Business Review*, July/August 1996, 100.

21. Chart, "The Making of a Giant," *New York Times*, October 8, 1998, C6.

22. Quoted in Margie Wylie, "Nader calls MS 'Uniquely Ruthless,'" *CNet*, November 14, 1997, at www.news.com/News/Item/0,4,16378,00.html.

23. James Gleick, "Control Freaks," *New York Times*, July 19, 1998, 18.

24. Steve Lohr, "'Browser War' Limits Access to Web Sites," *New York Times*, December 8, 1997, D1.

25. The U.S. Department of Justice's antitrust complaint against Microsoft alleged that the company's choice of partners for the channel bar amounted to an anticompetitive exclusionary practice. The Justice Department claimed specifically that, in order to get placed on the channel bar, Microsoft's partners had to agree not to do business with Microsoft's main competitors, including Netscape. See Complaint, *United States v. Microsoft Corp.*, May 18, 1998, D.D.C. (No. 98–1232), paragraphs 87–92.

    Even Robert Bork, the noted conservative legal scholar and critic of antitrust law, has expressed his discomfort with the fact that Microsoft, as he puts it, "controls what the consumer sees." Robert H. Bork, "What Antitrust is All About," *New York Times*, May 4, 1998, A23.

26. See Project to Promote Competition and Innovation in Digital Age, "At the Crossroads of Choice," April 20, 1998 , at www.procompetition.org/research; Larry G. Locke, Note, "Flying the Unfriendly Skies: The Legal Fallout Over the Use of Computerized Reservation Systems as a Competitive Weapon in the Airline Industry," *Harvard Journal of Law and Technology* 2 (1989): 219 (describing federal Civil Aeronautics Board's response to screen bias in computerized airline ticketing).

27. See Doreen Carvajal, "For Sale: Amazon.com's Recommendations to Readers," *New York Times*, February 8, 1999, A1; Doreen Carvajal, "Amazon.com Plans to Revise Its Ad Program," *New York Times*, February 10, 1999, C1.

28. See Amy Harmon, "Yahoo Offers to Expedite Its Site Reviews, for a Fee," *New York Times*, February 12, 1999, C19; "GoTo Sells Positions," Search Engine Watch, March 3, 1998, at searchenginewatch.internet.com/sereport/9803-goto.html.

29. Infoseek Press Release, "Infoseek Launches New Service Featuring Intelligent Channels," October 20, 1997.

30. "Kiss your browser goodbye," *Wired* magazine said in a cover story that grandly predicted the coming dominance of push media online. *Wired*, July 1997.

31. And there may be other benefits and costs to weigh. For example, in late 1998, news reports revealed that WebTV was "quietly using a system-polling feature that can extrapolate subscriber information from each of its 450,000 users to better serve advertisers." Karen J. Bannan, "WebTV is Watching You," *Interactive Week Online,* October 12, 1998, at www.zdnet.com.

CHAPTER 9: NARROWING OUR HORIZONS

1. Christopher Hitchens, "Norman Mailer: A Minority of One," *New Left Review*, March/April 1997, 115. These expressions are not uncommon: "With computers, individuals can be egomaniacs," says Tibor Kalman, editor and designer. Brad Wieners, "Color Him a Provocateur," *Wired*, December 1996, 258.

2. Turkle, *Life on the Screen,* 36. Michael Dertouzos, head of MIT's Laboratory for Computer Science, similarly uses an automotive analogy to describe the transition in the early 1980s from mainframe computers, which users had to share, to PCs: "Getting your own machine was like buying a car; you'd never again have to wait for the bus." Michael L. Dertouzos, *What Will Be: How the New World of Information Will Change Our Lives* (San Francisco: HarperEdge, 1997), 31. (Dertouzos's bus reference is an allusion to old time-sharing mainframe computers; before the advent of PCs, many people would be logged onto them at once, and users often had to wait their turn to get on.)

3. A search in the Nexis "News" database reveals 77 references to "information superhighway" in 1992, many of them mentioning Al Gore. The database also reveals a reference to "information superhighway" as early as 1983, though that use referred to regional fiber-optic link-ups, rather than a single, all-encompassing communications infrastructure. See William D. Marbach, "The Dazzle of Lasers," *Newsweek*, December 3, 1983, 36.

   Highway metaphors for electronic communication actually go back as far as the beginning of the twentieth century. In 1910, Theodore Vail, president of AT&T, described the deployment of copper telephone lines as "comprehensive, universal, interdependent, intercommunicating *like the highway system of the country*, extending from every door to every other door, affording electrical communication of every kind, from every one at every place to every one at every other place." Quoted in Pool, *Technologies of Freedom,* 29–30 (emphasis added).

4. Brand, *The Media Lab,* 37.

5. Ibid., 37–38.

6. Miller, *Civilizing Cyberspace,* 327 (quoting Herb Brody in *Technology Review*, April 1995).

7. A description of the ignore command for Internet Relay Chat can be found at www.newircusers.com/mirccmds.html. Leading email programs such as Eudora have sophisticated filters that allow users to screen out messages from certain individuals.

8. Like the bar code protocol for commercial packaging (that is, the space on the

package where the bar code is placed, as well as its size, shape, and dimensions), PICS is a way of standardizing how digital content such as web sites will be labeled.

9. See Lawrence Lessig, "Tyranny in the Infrastructure," *Wired*, May 1997, 96.

10. Cass Sunstein, "The First Amendment in Cyberspace," *Yale Law Journal* 104 (1995): 1757, 1786. Lawrence Lessig makes the same point: "The problem with systems like this is their perfection. The problem is that they permit people to tune out." Lessig, "The Law of the Horse." Total filtering is evident in technological contexts other than the Internet. Caller ID, for example, allows us to control our telephone environment in wholly new ways—screening with nearly absolute accuracy the calls we will take.

11. The notion of privatizing leisure time has been associated with television. See, for example, Robert Putnam, "Bowling Alone," *Journal of Democracy* 6(1) (January 1995): 65–78. Yet the personalization made possible by the Net suggests a more complete privatization of experience. Individuals may forgo not only close associations with family members and neighbors, but the "wider and shallower" bonds that the watching of common television programs might provide.

12. Leon Festinger, *A Theory of Cognitive Dissonance* (Stanford, Calif.: Stanford University Press, 1957).

13. Ibid., 3.

14. Ibid., 22.

15. Marylene Cloitre, "Avoidance of Emotional Processing: A Cognitive Science Perspective," in Dan J. Stein and Jeffrey E. Young, eds., *Cognitive Science and Clinical Disorders* (San Diego: Academic Press, 1992), 24.

16. Ibid., 21.

17. Sometimes this notion has been stated so forcefully as to seem overwrought, as when a presidential advisory committee stated in 1962 that if "a new piece of information would weaken the existing structure of [individuals'] ideas and emotions, it will be shunned." Quoted in William J. McGuire, "Selective Exposure: A Summing Up," in Robert P. Abelson et al., eds., *Theories of Cognitive Consistency: A Sourcebook* (Chicago: Rand McNally, 1972), 797, 798. The selective exposure hypothesis is also not without its critics. But this concept of avoidance has a genealogy that long predates modern psychology. As one recent text notes, "the idea that people tend to approach and attend information that upholds their attitudes and beliefs but avoid or pay little attention to conflicting information can be traced to William James, and before him, to Francis Bacon." Alice H. Eagly and Shelly Chaiken, *The Psychology of Attitudes* (Fort Worth, Tex.: Harcourt, Brace, Jovanovich College Publishers, 1993), 591; see also Editor's introduction, "Selective Exposure to Information," in Abelson et al., *Theories of Cognitive Consistency*, 769, noting that selective exposure theory arose "long before" consistency theories such as cognitive dissonance.

18. Elihu Katz, "Selectivity in Exposure to Mass Communications," in Abelson et al., *Theories of Cognitive Consistency*, 788, 793.

19. J. Mills, "Avoidance of Dissonant Information," *Journal of Personality and Social Psychology* 2 (1965): 589–93.

20. Festinger, *A Theory of Cognitive Dissonance*, 4.

21. Ibid., 20.

22. Stephanie Miles, "IBM Says PC On Its Last Legs," *CNet*, January 14, 1999 (quoting Paul Horn of IBM predicting the rise of "pervasive computing"), at www.news.com/News/Item/0%2C4%2C30954%2C00.html; Gary Chapman, "The Future Lies Beyond the Box," *Los Angeles Times*, January 4, 1999 (speaking of "ubiquitous computing").

23. John Perry Barlow, remarks at Digital China/Harvard conference, Cambridge, Mass., March 6, 1998.

24. Skeptics of selective avoidance theory might say that a person who believed homosexuality was wrong might be *more* inclined to read an article or listen to a conversation about gay life, if only to shore up his preexisting belief. But it is unlikely that this is true except for those who are actively engaged in a certain issue or area, and would therefore be able to make special use of even nonsupportive information. See Elihu Katz, "Selectivity in Exposure to Mass Communications," 788, 792 (noting that usefulness of information may cancel out the effect of its being nonsupportive). An active anti-abortion activist, for example, might try to expose herself to pro-choice messages, if only to know her opposition better. More routinely, though, research suggests that we seek out information that bolsters what we already believe.

25. Jeffrey Abramson, "The Internet and Community," *The Emerging Internet* (Queenstown, Md.: Aspen Institute, Institute for Information Studies, 1998), 59, 76.

26. This is what philosophers might call a "future choice" problem. At a certain time (*t1*), we may be inclined to make certain decisions about our interests and desires. But there is little way for us to know if those choices will remain valid at a later time (*t2*). Personalization and agent software can exacerbate this problem by committing us strongly to our *t1* choices throughout *t2, t3,* and so on.

    The decisions might not even all be our own. One of the most celebrated forms of agent software, known as collaborative filtering, makes recommendations to consumers—of music and books, for example—based on aggregation of preferences and a law of averages. If I like the same rock group and jazz quartet as Alice, then the software presumes that I will also enjoy the opera company that Alice likes. It therefore recommends the opera company to me. Why? Because, based on the accumulated feedback, a high percentage of the people who liked Alice's preferred rock group and jazz quartet also liked her preferred opera company. The problem is that there is a decent chance that I, the new user, am in the sizeable minority that likes the same rock group and jazz quartet as Alice but not her preferred opera company. Yet given the software's pretenses of scientific accuracy, I may be lulled into believing that the choice presented to me is the one I want.

Chapter 10: A Fraying Net

1. Ralph Reed, interview with author, August 18, 1997 (Aspen, Colorado).

2. Benedict Anderson, *Imagined Communities: Reflections on the Origin and Spread of Nationalism* (London: Verso, 1983).

3. See Shenk, *Data Smog,* 121 ("when the world becomes so profoundly splintered into distinct consumer tribes, humankind begins to lose the most valuable thing it has ever had: common information and shared understanding").

4. David Shaw, "Revolution in Cyberspace; Digital Age Poses the Riddle of Dividing or Uniting Society," *Los Angeles Times*, June 15, 1997, A26.

5. Consider the dilemma of the Burmese democracy activist discussed in Part One,

Htun Aung Gyaw. Htun uses the Net to stay in touch with fellow dissidents who are fighting the repressive government of Burma. While cyberspace would seem to be a big boon to his life, the *New York Times* profile of him also showed the way in which his online obligations impinged upon other commitments in his life. The *Times* reporter, William Glaberson, wrote: "His wife said she supports his political work but sometimes, when he turns on the computer, she finds herself growing outraged. 'He should do that,' she said. 'I agree with that. But I and my children were away from him for six years. And when we arrived, he had no time for us.'" And Glaberson noted that Htun was distracted at work, as well: "Even some of those who say they are admirers say Mr. Htun is sometimes so distracted by his involvement in the movement as it passes through his computer screen that he is unable to do what is expected of him." William Glaberson, "A Guerilla War on the Internet," *New York Times*, April 8, 1997, B1.

6. Robert Putnam, "Bowling Alone," *Journal of Democracy* 6(1) (January 1995), 65–78.

7. Theodore Roszak, *The Cult of Information: A Neo-Luddite Treatise on High Tech, Artificial Intelligence, and the True Art of Thinking* (Berkeley: University of California Press, 1994), 169.

8. Quoted in Levy, *Hackers*, 289–290.

9. Even Ithiel de Sola Pool, the influential scholar who in the 1970s and 1980s predicted much of the positive potential of interactive technology, voiced serious concerns about personalization in a book published posthumously in 1990. "Technological trends," he concluded in the last lines of the book, "will promote individualism and will make it harder, not easier, to govern and organize a coherent society." Ithiel de Sola Pool, *Technologies Without Boundaries* (Cambridge, Mass.: Harvard University Press, 1990), 261–262.

10. Some cyber-romantics seem to approve of this development. See Davidson and Rees-Mogg, *The Sovereign Individual,* 284 ("As individuals themselves begin to serve as their own news editors, selecting what topics and news stories are of interest, it is far less likely that they will choose to indoctrinate themselves in the urgencies of sacrifice for the nation-state").

11. Quoted in Lawrence K. Grossman, *The Electronic Republic: Reshaping Democracy in the Information Age* (New York: Twentieth Century Fund/Viking Press, 1995), 156. On the theme of social fragmentation generally, see Todd Gitlin, *The Twilight of Common Dreams: Why America is Wracked by Culture Wars* (New York: Metropolitan Books, 1995), and Arthur M. Schlesinger Jr., *The Disuniting of America* (New York: W. W. Norton and Company, 1992).

12. Aboriginal hunter-gatherers in Australia, for example, engage in a wide variety of cultural practices that help them to endure stressful environmental conditions such as drought. See Richard A. Gould, "Rock Pools and Desert Dances," *Natural History,* March 1984, 63, and "To Have and Have Not: The Ecology of Sharing Among Hunter-Gatherers," in *Resource Managers: North American and Australian Hunter-gatherers*, ed. Nancy M. Williams and Eugene S. Hunn (Boulder, Colo.: Westview Press, 1981).

13. Esther Dyson, *Release 2.0: A Design for Living in the Digital Age* (New York: Broadway Books, 1997), 8.

14. Robert Kraut, Michael Patterson, Vicki Lundmark, and Sara Kiesler, "Internet Paradox: A Social Technology that Reduces Social Involvement and Psychological Well-being?" *American Psychologist* 53(9) (1998), 1017–1031. For one of the bet-

ter responses to the study, see Scott Rosenberg, "Sad and Lonely in Cyberspace," *Salon*, September 3, 1998, at www.salonmagazine.com.

15. Daniel Lerner, *The Passing of Traditional Society* (Glencoe, Ill.: Free Press, 1958).

16. Pool, *Technologies Without Boundaries*, 15.

CHAPTER 11: FREEDOM FROM SPEECH

1. Spaces such as streets and parks "have immemorially been held in trust for the use of the public and, time out of mind, have been used for purposes of assembly, communicating thoughts between citizens, and discussing public questions," the Supreme Court said in *Hague v. CIO*, 307 U.S. 496, 515 (1939). The Court has held, more precisely, that content-based government restrictions on speech in the public forum are unconstitutional under the First Amendment unless supported by a narrowly tailored law that serves a compelling state interest. See, for example, *Perry Education Association v. Perry Local Educators' Association*, 460 U.S. 37, 45–46 (1983); *Rosenberger v. Rectors and Visitors of the University of Virginia*, 515 U.S. 819, 828–830 (1995). The First Amendment, in limited circumstances, has also been held to prevent powerful private actors from inhibiting speech in public forums. See *Marsh v. Alabama*, 326 U.S. 501 (1946).

2. See, for example, *International Society for Krishna Consciousness v. Lee*, 505 U.S. 672 (1992) (restricting free speech rights of individuals in a public airport); *United States v. Kokinda*, 497 U.S. 720 (1990) (restricting free speech rights on a public sidewalk adjacent to a post office); *Lloyd Corp. v. Tanner*, 407 U.S. 551 (1972) (restricting free speech rights in a shopping mall). For an excellent treatment of these issues, see Owen M. Fiss, *The Irony of Free Speech* (Cambridge, Mass.: Harvard Univ. Press, 1996), and Owen M. Fiss, *Liberalism Divided: Freedom of Speech and the Many Uses of State Power* (Boulder, Colo.: Westview Press, 1996).

3. The Supreme Court has made clear that, for free-speech purposes, it is always better for citizens to bear the burden of avoiding unwanted speech than it is to silence that speech. See, for example, *Madsen v. Women's Health Center*, 512 U.S. 753, 773 (1994); *Erznoznik v. City of Jacksonville*, 422 U.S. 205, 209 (1975).

4. Fortunately, American courts have recognized that the right to free speech may also imply the right to speak in certain venues and to certain audiences. See, for example, *Schneider v. State*, 308 U.S. 147, 163 (1939) ("One is not to have the exercise of his liberty of expression in appropriate places abridged on the plea that it may be exercised in some other place."); *City of Ladue v. Gilleo*, 512 U.S. 43 (1994) (striking down ordinance preventing citizens from posting signs on personal property on grounds that citizens had right to speak to particular audience of neighbors); *Bery v. City of New York*, 97 F.3d 689 (2d Cir. 1996), cert. denied, 117 S. Ct. 2408 (U.S. 1997) (striking down licensing scheme for street artists on grounds that artists had right to display visual work on streets of New York as opposed to in their homes or in galleries).

5. This is Justice William Brennan's renowned formulation, from *New York Times v. Sullivan*, 376 U.S. 254, 270 (1964). For more on this, see Fiss, *Liberalism Divided*, Chapter 2.

6. See Bruce Ackerman, *Social Justice in the Liberal State* (New Haven: Yale University Press, 1980).

7. *FCC v. Pacifica Foundation,* 438 U.S. 726, 749 n.27 (1978).
8. *Rowan v. United States Post Office Department,* 397 U.S. 728, 738 (1970).
9. See *Abrams v. United States,* 250 U.S. 616, 630 (1919) (Holmes, J., dissenting) (advocating "free trade in ideas—that the best test of truth is the power of the thought to get itself accepted in the competition of the market").
10. Gates, November 1995 Public Lecture.
11. See, for example, Eugene Volokh, "Cheap Speech and What It Will Do," *Yale Law Journal* 104 (1995): 1805.
12. In early 1999, for example, a start-up company called Free-PC (www.free-pc.com) offered 10,000 free computers and Internet connections to individuals who would agree to watch selected onscreen advertisements on their new computers, regardless of whether they were connected to the Internet. Close to a million people apparently tried to sign up for the deal. See Matt Richtel, "Plan for Free PC's Has a Few Attachments," *New York Times,* February 8, 1999, C8; Anita Hamilton, "Your Technology: Free PCs, for a Price," *Time,* February 22, 1999, 98.
13. For example, Roger Tamraz, an oil financier, gave $177,000 to Democrats in 1995 and 1996 so that he could gain access to President Clinton through "coffees" with the president. Anne Fares, "Unfolding Story Swelling Like a Sponge," *Washington Post,* April 6, 1997, A16.
14. Gates, November 1995 Public Lecture.
15. See, for example, "This Space Available on AOL Starting for a Mere $25,000," *New York Times,* October 1, 1998, G3.
16. See, for example, www.aolsucks.com.
17. Quoted in Michael Marriot, "Amplifying Voices for Human Rights," *New York Times,* February 26, 1998, G14.

CHAPTER 12: THE DRUDGE FACTOR

1. See *Blumenthal v. Drudge,* 992 F. Supp. 44 (D.D.C. 1998). See also Andrew Hearst, "Is AOL Responsible for its Hip Shooter's Bullets?" *Columbia Journalism Review,* November/December 1997, 14.
2. Steven Brill, "Pressgate," *Brill's Content,* July/August 1998, 123, 129.
3. Janny Scott, "A Media Race Enters Waters Still Uncharted," *New York Times,* February 1, 1998, A1, A19. See also Brill, "Pressgate."
4. Drudge himself admits this: "I realized early on, it is easier to sleep at night if you can say at every step that you reported the truth as you knew it." Matt Drudge, "The Media Should Apologize," September 10, 1998, text of address to the Wednesday Morning Club, at www.frontpagemag.com/archives/drudge/wmcspeech.htm.
5. Drudge has demonstrated the danger of his method. For a few days in January 1999, he sent out breathless reports about a tabloid's investigation of what turned out to be a baseless claim that Clinton had fathered a child out of wedlock. Drudge reported that the child, now a teenager, was having his DNA tested to see if it matched Clinton's. He had hyped the story so much that he even felt obliged to sensationalize the revelation that there was no story:

> White House DNA Chase: Teen Doing 'Well' After News of 'No Match'
> **Exclusive**

He had been told all of his life by his mother that Bill Clinton was his father, but late this week, 13-year old Danny Williams of Arkansas learned the truth: He is not.

A stunning DNA showdown came to a dramatic conclusion this weekend when it was learned that STAR magazine was in posession [sic] of lab results—results that ruled out Bill Clinton as the father!

6. Christopher Hanson, "The Dark Side of Online Scoops," *Columbia Journalism Review*, May/June 1997, 17.
7. David Jackson, "Source Affirms Clinton Affair," *Dallas Morning News*, January 26, 1998, A1; David Jackson, "Information for Story Inaccurate, Source Says," *Dallas Morning News*, January 27, 1998, A1.
8. See Howard Kurtz, "Wall Street Journal Story Is Rushed Onto the Web," *Washington Post*, February 5, 1998, A12; Janny Scott, "Internet Story Revives Questions on Standards," *New York Times*, February 6, 1998, A20.
9. Quoted in Lee Marshall, "The World According to Eco," *Wired*, March 1997, 145, 148.
10. See Amy Harmon, "Diana Photo Restarts Debate Over Lack of Restrictions on Internet Postings," *New York Times,* September 22, 1997, D9.
11. See Todd S. Purdum, "The Dangers of Dishing Dirt in Cyberspace," *New York Times*, Week in Review, August 17, 1997, 3.
12. Organizations that monitor human rights depend heavily on accurate reporting because it bears directly on the validity of their claims about abusive practices. Because of the global nature of their work, these groups increasingly rely on the Net (as noted earlier) for their watchdog activities. But because partisans may alter the facts to serve their own political ends, these organizations cannot vouch for every act of brutality or torture that someone reports by email. Says one organizer, "We don't post stuff from people we don't know." Quoted in A. Lin Neumann, "The Resistance Network," *Wired*, January 1996, 108, 112.
13. Grassroots publishers also have a legal duty to refrain from libelous statements. Libel law, though, will likely evolve as the Net emerges, because (among other things) it will be easier for individuals to respond to false statements. See Andrew L. Shapiro, "Drudge Match," *Wired*, April 1998, 89.
14. This, in fact, is what we should glean from Pierre Salinger's supposedly notorious run-in with the Internet. In the fall of 1996, the former ABC reporter and spokesman for President Kennedy declared that he had obtained a classified government document showing that the crash of TWA flight 800 earlier that summer was inadvertently caused by a U.S. Navy missile. As the claim was refuted by experts and investigators, many observers emphasized that Salinger's bogus document came from the Internet, citing it as an example of how unreliable information is online. What they conveniently ignored, however, was Salinger's undisputed statement that he initially had no contact with the Internet and no idea the document had originated there. Rather, he got it from an acquaintance who simply conned him into believing—quite easily, it seems—that it was a top-secret government report. It was, in other words, less a case of cyber deception than a run-of-the-mill hoax. As gullible as Salinger may have been, moreover, it was the old media that gave the story legs. Rumors about the "friendly fire" theory had been circulating for months on the Net, with no real effect. Indeed, care-

ful rebuttals of the "friendly fire" theory appeared in discussion forums on the Internet even before Salinger made his claim. See "flight-800" listserv archive, at www.lsoft.com/listserv.stm.

15. Drudge, "The Media Should Apologize."

Chapter 13: Shopper's Heaven?

1. Gates, *The Road Ahead,* 181.

2. The numbers are, of course, difficult to quantify precisely. But, according to federal labor statistics, wholesalers and retailers make up roughly 20 percent of all employed individuals in the United States. And nearly a third of all other employed Americans are in "professions and related services"—law, education, health care, social services—where some disintermediation will likely occur. See U.S. Census Bureau, *Statistical Abstract of the United States: 1998* (1998), 421.

3. Neil Munro, "Industrial Policy for the Information Age," *National Journal,* March 28, 1998, 698.

4. Software designers, in fact, are working on computer programs that are actually meant to provide nuanced legal judgment. The Virtual Patent Advisor, for example, is a program that purportedly knows legal rules and can apply them to the facts of a patent query. See Geanne Rosenberg, "A Lawyer Minus the Briefcase," *New York Times,* November 10, 1997, D7.

5. Richard E. Sclove, "The Democratic Uses of Technology," *Thought and Action: The NEA Higher Education Journal* 14(1) (Spring 1998), 9, 12, available at www.nea.org/he/tanda.html.

6. Another possible scenario is that large bookstores could see their competitive advantage—selling high volume at low prices—undermined by online booksellers. And this might unexpectedly give small bookstores a chance to rebound and focus on what they do best: provide service and atmosphere.

7. Statement by Chairman Arthur Levitt Concerning On-line Trading, January 27, 1999.

8. Rick Berry, director of equity research at J. P. Turner and Co., quoted in Richard Melville, "As 'Net Fever Mounts, Some Fear for Market Health," Reuters, December 29, 1998.

9. William Hambrecht, quoted in Denise Caruso, "The Online Day-Trader Phenomenon," *New York Times,* December 14, 1998, C3.

10. Phil Feigin, a Colorado securities regulator, quoted in Sana Siwolop, "Online Investors Chase Every Blip," *New York Times,* October 1, 1998, G1.

11. Wharton School professor Jeremy Siegel, quoted in David Barboza, "Online Traders: Older, Wiser and Richer," *New York Times,* December 20, 1998, sec. 3, 1.

12. See Gates, *The Road Ahead,* 180.

13. See 15 U.S.C. sec. 77k, 77l(2) (1998).

14. See *Ryan v. Progressive Grocery Stores,* 255 N.Y. 388, 395 (N.Y. 1931).

15. The justifications for this rule are well explained in the Restatement of Torts. See Restatement of Torts sec. 402A, comment f ("The basis for the rule is the ancient one of the special responsibility for the safety of the public undertaken by one who enters into the business of supplying human beings with products which may endanger the safety of their persons and property").

16. *See* Restatement of Torts sec. 402A, illustration 1.

17. An exception is that Amazon must collect sales tax for orders shipped to customers in Washington state, where the company is based. According to the Supreme Court's ruling in *Quill Corp. v. North Dakota*, 504 U.S. 298 (1992), the Constitution prohibits states from requiring remote-sales vendors (whether they do business by mail, phone, or Internet) to collect sales taxes unless the vendor has a substantial nexus with the state in question—for example, if it is located there.

18. In Connecticut, 1 in 100 people reportedly pay use taxes and in New Jersey 1 in 500 do. David Cay Johnston, "The Old Tax Dodge," *New York Times*, April 15, 1998, D1.

19. On the other hand, Internet advocates have expressed concern about how competing claims from multiple taxing authorities will be sorted out; they fear that too much taxation will inhibit growth of the Net and electronic commerce. A favorite rhetorical device of this crowd is to point out that there are at least 30,000 taxing authorities in the world. The Internet Tax Freedom Act of 1998 imposed a temporary moratorium on new taxes on e-commerce; existing sales taxes are allowed.

20. Lisa M. Bowman, "California joins the Internet tax-ban bandwagon," ZDNet News, August 24, 1998, at www.zdnet.com. See also Center on Budget and Policy Priorities, "A Federal 'Moratorium' on Internet Commerce Taxes Would Erode State and Local Revenues and Shift Burdens to Lower-Income Households," May 11, 1998, at www.cbpp.org/512webtax.htm.

CHAPTER 14: PUSH-BUTTON POLITICS

1. See www.parolewatch.org.

2. These statements were scheduled to be added to the site in early 1999.

3. U.S. Census Bureau, *Statistical Abstract of the United States: 1998* (1998) 276–302.

4. Walter Lippmann, *The Public Philosophy* (New York: New American Library, 1955), 19.

5. See, for example, Lawrence Lessig, "The Spam Wars," *Industry Standard,* January 11–18, 1998, 16. To be sure, there will be positive corollaries. Citizens will be able to use networked video cameras and online archives to track crime in their communities and even to keep an eye on potentially abusive police. Yet it may be all too easy for a few to cross the line and take the law into their own hands, simply because they have the ability to do so.

6. Cronin, *Direct Democracy,* 24–25.

7. Even ballot initiatives and referenda, which are the closest thing we have now to direct democracy, apparently often prompt confusion and "ballot fatigue" among voters. See ibid., 68, 198. Insisting on deliberation before decisionmaking, it should be clear, is not about mistrusting the masses. It is about mistrusting anyone, including the most seasoned and erudite representative, who decides on impulse how state power shall be used. Thoughtful proponents of populist democracy recognize that deliberation cannot be forsaken. As Benjamin Barber says, "Soliciting instant votes on every conceivable issue from an otherwise uninformed audience that has neither deliberated nor debated an issue would be the death of democracy." Benjamin Barber, *Strong Democracy: Participatory Politics for a New Age* (Berkeley: University of California Press, 1984), 290.

8. One exception is James Fishkin, a political scientist who has devised what he calls a deliberative poll. The poll begins with a long period, often stretching over a few days, during which respondents are exposed to different views on an issue, given a chance to debate, and then ultimately asked for their views. See James S. Fishkin, *The Voice of the People: Public Opinion and Democracy* (New Haven: Yale University Press, 1995).

9. See *Federalist* No. 10. It's worth noting that we don't have to glorify the entire political vision of the Founders in order to make the case against direct democracy. Most glaringly, the Constitution did not originally guarantee the right to vote for women, blacks, or the poor; indeed, it condoned slavery and counted a slave as three-fifths of a person.

10. For example, on freedom of expression, see "Most Want Flag-Burning Ban," *Chicago Tribune*, June 11, 1997, 6 (81 percent prefer ban on flag burning); on prayer in schools, see Richard Benedetto, "Economy Shakes American Dream," *USA Today*, January 16, 1992, 5A (62 percent favor a constitutional amendment allowing prayer). And notably, many recent attempts to curb minority rights—such as affirmative action and immigrants' rights in California, and gay rights in Colorado— have come about as a result of ballot initiatives. For more on American attitudes toward civil liberties, see Herbert McClosky and Alida Brill, *Dimensions of Tolerance: What Americans Believe about Civil Liberties* (New York: Russell Sage Foundation/Basic Books, 1983), and Dennis Chong, "How People Think, Reason, and Feel about Rights and Liberties," *American Journal of Political Science*, 37 (August 1993): 867–899.

11. As historian Gordon Wood, author of *The Radicalism of the American Revolution*, says: "They saw the public interest as a transcendent thing that enlightened people would be able to see and promote. It wasn't just a question of adding up all the interests." Quoted in Robert Wright, "Hyper Democracy," *Time*, January 23, 1995, 15, 18. Madison's concerns about factionalism also imply a concern for minority rights: "When a majority is included in a faction, the form of popular government, on the other hand, enables it to sacrifice to its ruling passion or interest both the public good and the rights of other citizens." *Federalist* No. 10.

Chapter 15: Privacy for Sale

Unless otherwise cited, all quotes and factual assertions in this chapter are from Andrew L. Shapiro, "Privacy for Sale: Peddling Data on the Internet," *Nation,* June 23, 1997.

1. Indeed, software that allows users to personalize experience online is a major new source for data collectors, as individuals' detailed preferences about material they find interesting are collected and aggregated. Privacy experts estimate that the average American is profiled in at least twenty-five, and perhaps as many as one hundred, databases.

2. Federal News Service, "Remarks by President Bill Clinton at Commencement Ceremony, Morgan State University, Baltimore, Maryland," May 18, 1997. Eighty-one percent of Internet users are concerned about threats to their personal privacy while online, according to a June 1998 Louis Harris poll. See idt.net/~pab/pabsurve.htm.

3. Information Policy Committee, National Information Infrastructure Task Force, "Options for Promoting Privacy on the National Information Infrastructure," April 1997 draft, at www.iitf.nist.gov/ipc/privacy.htm.

4. Quoted in Jeri Clausing, "Proposed Standards Fail to Please Advocates of Online Privacy," *Cybertimes* (*New York Times*), June 2, 1998, at www.nytimes.com/library/tech/98/06/cyber/articles/02privacy.html.

5. The market for privacy might also affect the viability of anonymous information exchange. In a world of formally established privacy markets, individuals on the Net (and elsewhere) will presumably have to contract with vendors in an above-board fashion. As Pat Faley of the Direct Marketing Association sees it, "You're not going to be able to go to too many places if you want to be anonymous."

6. The site was changed after complaints from the Center for Media Education (www.cme.org).

7. Oscar H. Gandy Jr., "Legitimate Business Interest: No End in Sight? An Inquiry into the Status of Privacy in Cyberspace," *University of Chicago Legal Forum* (1996): 77, 127.

8. As David Brin argues, we should not assume that absolute privacy always works in society's favor. There are times when what we really want is mutual transparency—the ability to scrutinize the behavior of powerful figures just as much as they scrutinize us. Brin, *The Transparent Society.*

9. See Carl S. Kaplan, "Strict European Privacy Law Puts Pressure on U.S.," *Cybertimes (New York Times),* October 9, 1998.

10. *Olmstead v. United States,* 277 U.S. 438, 478 (1928) (Brandeis, J., dissenting). See also Samuel D. Warren and Louis D. Brandeis, "The Right to Privacy," *Harvard Law Review* 4 (1890): 193.

11. See Margaret Jane Radin, *Contested Commodities* (Cambridge, Mass.: Harvard University Press, 1996).

CHAPTER 16: MAPPING PRINCIPLES

1. See, for example, *Reno v. American Civil Liberties Union,* 521 U.S. 844 (1997) (ruling that the Internet is entitled to the full First Amendment protection accorded to printed speech, as opposed to the more limited protection granted to broadcast media).

2. See, for example, *FCC v. Pacifica, Reno v. ACLU.*

3. *Reno,* 521 U.S. at 893, 904.

4. A number of valuable critiques of filtering software have been produced, including Electronic Privacy Information Center, "Faulty Filters: How Content Filters Block Access to Kid-Friendly Information on the Internet," at www2.epic.org/reports/filter-report.html, and American Civil Liberties Union, "Fahrenheit 451.2: Is Cyberspace Burning? How Rating and Blocking Proposals May Torch Free Speech on the Internet," at www.aclu.org/issues/cyber/burning.html.

5. In 1998, Congress tried a more subtle form of code regulation, one that came closer to recognizing that there might be middlemen on the Net: It passed the Child Online Protection Act (COPA), a more narrowly tailored Internet indecency law, known informally as CDA 2, which applied only to commercial web sites (and used a more limited definition of what material was supposed to be restricted). At the time of this writing, COPA had been preliminarily enjoined by a federal

judge on the grounds that it is likely unconstitutional. See Pamela Mendels, "Setback for a Law Shielding Minors From Adult Web Sites," *New York Times*, February 2, 1999, A12.

Internet service providers, it should be noted, have been effectively immunized from liability for indecent content transmission by section 230 of the Communications Decency Act, which is still good law. 47 U.S.C. sec. 230(c)(1).

6. The white-list kid browser might also allow adult users to add web sites to the list of acceptable destinations.

7. Requiring Net users to take affirmative steps to establish their adult status—and, by implication, their desire to see more than kiddie fare—could present privacy problems. But these should be cured by allowing an individual to establish his adult status anonymously through a third party. That way no one would know he had downloaded an adult browser.

8. With the surveillance of trusted systems, there would also appear to be no wiggle room to let pass what would otherwise be *de minimis* uses.

CHAPTER 17: SHATTERING ILLUSIONS

1. See Katie Hafner, "Net Presence: If No One Sees a Web Site, Is It Really There?" *New York Times*, January 7, 1999, G5.

2. A number of states have alleged that Microsoft's bundling of various software applications—Word, Excel, PowerPoint, and so on—in a single software suite is anticompetitive. See Joel Brinkley, "Microsoft Case Is Set for Trial In September," *New York Times*, May 23, 1998, D1.

3. See Andrew L. Shapiro, "aol.mergergame.com," *Nation*, December 21, 1998, 7.

4. "It is a sad but telling fact that in the high-tech field, virtually no business plan will be financed today without a convincing answer to the question of what is to be done about competition from Microsoft," says software entrepreneur Mitchell Kapor in "High-Tech Hypocrisy About Government," *New York Times*, May 26, 1998, A21.

5. The main antitrust trial was still underway as of this writing.

6. See Amy Harmon and John Markoff, "Internal Memo Shows Microsoft Executives' Concern Over Free Software," *New York Times*, November 3, 1998, C8. Copies of memos expressing Microsoft's concern about open source are available at www.opensource.org/halloween.html.

7. Two-thirds of experienced Internet users surveyed, for example, said they would not buy an early WebTV model precisely because of the way the product inhibited choice by not allowing users to choose their own Internet service providers. Diane Crispell, "The Internet on TV," *American Demographics*, May 1997, 32.

8. Seligman and Miller, "The Psychology of Power," in *Choice and Perceived Control*, 347, 348.

9. Erich Fromm, *Escape from Freedom* (New York: Rinehart, 1941), 253.

CHAPTER 18: IN DEFENSE OF MIDDLEMEN

1. Kelly, *Out of Control*, 127.

2. Nicki Grauso, quoted in Lee Marshall, "The Berlusconi of the Net," *Wired*, January 1996, 82. A similar sentiment was expressed by Ted Leonsis of America On-

line: "In the near future, everyone will have access to all the information they need
to make their own decisions. So who needs the media to deliver content? I hate
to say it, but I think the media are in a death spiral." Jack O'Dwyer, "Revolution
Has Started," *Jack O'Dwyer's Newsletter*, February 22, 1995, 3, quoted in Shenk,
Data Smog, 165.

3. See, for example, the incidents cited at the end of Chapter 12.

4. As David Shenk says, "If not journalists, who else will expose medical frauds and care-
less doctors? Who else will hold politicians to their promises? Who else will examine
the design, intent, and honesty of advertising? Who else will monitor the link between campaign contributions and political favors?" Shenk, *Data Smog*, 166.

5. See Kurt Andersen, "The Age of Unreason," *New Yorker*, February 3, 1997, 40, 42.

6. Christopher Lasch made the argument against objectivity eloquently in "Journalism, Publicity, and the Lost Art of Argument," a 1990 essay in which he argued
that democracy needs more opinionated debate. The essay appears as Chapter 9
in a collected volume, Christopher Lasch, *The Revolt of the Elites and the Betrayal
of Democracy* (New York: W. W. Norton, 1996).

7. Quoted in J. D. Lasica, "What Journalism Can Bring to the Net," *American Journalism Review/NewsLink*, July 7–13, 1998, at www.newslink.org/ajrjd11.html.

8. Media entrepreneur Steve Brill makes the point compellingly in this vignette:

> In 1969, many of us woke one morning to find a front-page story from a re-
> porter named Seymour Hersh about American soldiers massacring women
> and children in Vietnam. Had we been hooked on the wonders of the In-
> ternet in those days, none of us readers would have woken up that morning
> and said 'Gee, I wonder whether Seymour Hersh has any information about
> Americans massacring women and children in Vietnam. Why don't I do a
> search?' This was information I didn't know I needed to know or wanted to
> know on that morning in 1969. Rather I paid the editor and the reporter,
> with my twenty-five cents for a newspaper, to tell me what they thought I
> should know. And I got my money's worth.

"For Brill, the Thrill is Print," *Columbia Journalism Review* (May/June 1997): 18.

9. Hibbitts's article, appropriately enough, appeared on the web first and was later
republished in the *New York University Law Review*. Bernard J. Hibbitts, "Last
Writes? Reassessing the Law Review in the Age of Cyberspace," *New York University Law Review* 71 (1996): 615.

10. Variations on this theme are occurring. See, for example, the Social Science Research Network online, at www.ssrn.com, which publishes abstracts of academic
papers and papers in full.

11. Madison's classic argument for representative democracy over direct democracy
is found in the *Federalist* No. 10.

12. Constitutional objections to such a reform plan are unwarranted in light of the
danger that a money-driven campaign system poses to our democracy. See, for example, E. Joshua Rosenkranz, *Buckley Stops Here: Loosening the Judicial Stranglehold on Campaign Finance Reform* (New York: Twentieth Century Fund, 1998).

13. The Berkman Center for Internet and Society is engaged in a preliminary project
with Professor Fishkin to explore deliberative polling online. See cyber.law.har-
vard.edu/9–10mtg/idp.html.

14. Some efforts to remedy this are already underway. They include the Cyberspace Law Institute's Deliberation Project. See www.cli.org.

15. Cronin, *Direct Democracy,* 38–40. An inspiring example of citizens using the Net to educate one another is the web site moveon.org, which urged Congress to censure, rather than impeach, President Clinton and to "move on" to the truly pressing business of the nation.

16. California and half a dozen other states, for example, refused to comply with the federal National Voter Registration Act of 1993 (better known as the motor voter law, because it allows citizens to register to vote when obtaining a driver's license). The states unsuccessfully sued the federal government, claiming that the law interfered with states' rights and imposed an unauthorized financial burden on them. See "Court Rejects Attack On 'Motor Voter' Law," *Legal Intelligencer,* January 23, 1996, 4.

17. Costa Rica, in fact, has already stated its intent to move completely to electronic balloting by the year 2002, using computer terminals in schools to prevent fraud. Global Internet Liberty Campaign, "Regardless of Frontiers," at www.gilc.org/speech/report.

18. Cokie Roberts and Steve Roberts, "Internet Could Become a Threat To Representative Government," United Features syndicate, April 5, 1997.

CHAPTER 19: IN DEFENSE OF ACCIDENTS

1. Ted Gup, "The End of Serendipity," *The Chronicle of Higher Education,* November 21, 1997, A52.

2. There may be, more generally, something unique about the way secular Westerners approach questions of power and control. To a Buddhist, for example, individual control may have much to do with self-denial, simplicity, and obeisance, whereas the control that technology makes possible is more about a lack of constraints, endless options, and self-indulgence. Yet it would be wrong, even in our context, to allow personal control to be equated solely with an absence of limits.

   Some proponents of this view even claim that personal empowerment is not really such a fundamental human need and that "Westerners have been conned into a need for control." Seligman and Miller, "The Psychology of Power" in *Choice and Perceived Control,* 347, 355. What is important, they say, is not so much personal dominion but predictability. And they add that while individual control is one way to cure unpredictability, there are other ways, too, like submitting to some guiding authority, whether human or divine.

3. *Reno,* 521 U.S. at 887.

4. Quoted in Robert Wright, "The Man Who Invented the Web," *Time,* May 19, 1997, 64, 68.

5. See Joseph Turow, *Breaking Up America: Advertisers and the New Media World* (Chicago: University of Chicago Press, 1997); Todd Gitlin, "Welcome to Nomadicity: We're All Alone Now," *New York Observer,* June 9, 1997.

6. Jacques Ellul, "The Power of Technique and the Ethics of Non-Power," in *The Myths of Information: Technology and Postindustrial Culture,* Kathleen Woodward, ed. (Madison, Wis.: Coda Press, 1980), 242, 245. "We need to recognize what's uncontrollable," says Judith Viorst. "We need to recognize when surrender makes sense." Viorst, *Imperfect Control,* 302.

7. Michael Hart, founder of Project Gutenberg, a nonprofit project that is making classic books available online for free, puts a similar spin on the issue. "Some people say, 'I am the most powerful because I have the most power,'" says Hart. "I say, 'I am the most powerful because I give the most power away.'" Denise Hamilton, "Hart of the Gutenberg Galaxy," *Wired*, February 1997, 108, 115.

8. Software designer Ellen Ullman, in an amusing critique of Microsoft's My Computer desktop icon, suggests that the problem with personalization may have something to do with the way it has been pitched to us. "*My Computer*. I've always hated this icon—its insulting, infantilizing tone," she says. "Even if you change the name, the damage is done: It's how you've been encouraged to think of the system. My Computer. My Documents. Baby names. My world, mine, mine, mine." Ellen Ullman, "The Dumbing Down of Programming," *Salon*, May 12, 1998, at www.salonmagazine.com/21st/feature/1998/05/cov_12feature2.html.

9. Gerald Marzorati, "How the Album Got Played Out," *New York Times Magazine*, February 22, 1998, sec. 6, 36, 38.

10. Consider this eloquent and unusual passage from a *New York Times* editorial describing the joys of being a pedestrian in a big city:

> Sometimes a curiosity emerges—perhaps the weeping anger of a man shouting at a delivery van in Times Square—and the day is instantly diverted from its ordinary course.
>
> Onlookers gather, and as you edge past them you can feel the nearly unavoidable temptation to look over the shoulder in front of you to see what's going on. You watch for a while and then walk on, and only when you pick up the thread of your thoughts once more do you realize how much of yourself you surrendered . . . .

Editorial, "Ricky Jay is Back in Town," *New York Times*, January 17, 1998, A12.

11. "In one's family or village, even in one's church or political party, there is always the challenge of coping with those whose opinions, attitudes, and interests differ radically from one's own," author Derek Bickerton notes. "In cyberspace, there's the seductive appeal of being able to confine oneself to meeting only those who share some obsession, of hanging out in chat rooms with virtual clones of oneself." Derek Bickerton, "Digital Dreams," *New York Times Book Review*, November 30, 1997, 6 (reviewing Esther Dyson, *Release 2.0: A Design for Living in the Digital Age*).

12. The quote is from James Gleick, *Chaos: Making a New Science* (New York: Penguin Books, 1987), 15, an excellent introduction to chaos theory.

13. Quoted in Brin, *The Transparent Society*, 50.

14. Quoted in Samuel G. Freedman, "Yeshivish at Yale," *New York Times Magazine*, May 24, 1998, 32, 34.

15. 929 F. Supp. 824, 883 (E.D. Pa. 1996).

16. "President Clinton's Message to Congress on the State of the Union," *New York Times,* February 5, 1997, 20.

17. Reno, 521 U.S. 897 (1997).

18. Specifically, the First Amendment prevents government from favoring certain speakers over others based on what they are saying. In a way, this is what Microsoft does when it prefers certain content and commerce links instead of others. It may

be able to deny speakers a license, then, at least in the sense of not giving them permission to appear on a privileged platform where they will be widely heard.

19. A silenced speaker like Paine might have a viable First Amendment claim if the entity silencing him were a powerful private actor such as a corporation that, by virtue of its dominance over a certain sphere, was exercising governmentlike powers. See *Marsh v. Alabama,* 326 U.S. 501 (1945) (ruling that a company town could not prevent distribution of printed materials on privately owned streets of business district) But if it were simply individuals exercising their filtering power in such a way that the overall result was that no one heard Paine, he would likely have no case at all under current First Amendment law.

20. PublicNet would fit well into a nascent movement in interface design called "web urbanism," which calls for the creation of online environments that preserve the serendipity of city spaces. "Perhaps the most important element of urban life is the chance encounter," says architect and web designer Eric Liftin. "The determinism of computer programming should not prevent accidents and unanticipated encounters. An interface must make accessible to the user not only requested results, but the unexpected as well." Eric Liftin, "Preliminary Notes on Web Urbanism," at http://www.mesh-arc.com/writings/arch-inter.html. See also Negroponte, *Being Digital,* 154; Johnson, *Interface Culture,* 63–64, 235.

   Web urbanism has much in common with an idea I suggested in an article a few years ago: namely that, among two spatial models of online interaction, which I called Cyburbia and Cyberkeley, we seemed to be choosing the order of the former over the occasional chaos of the latter. See Andrew L. Shapiro, "Street Corners in Cyberspace," *Nation,* July 3, 1995, 10.

21. Why this speech right applies equally to corporations and individuals is a curiosity that has confounded many scholars, but it is the law. See Fiss, *The Irony of Free Speech,* 3.

22. See *Associated Press v. United States,* 326 U.S. 1 (1945); *Turner Broadcasting System, Inc. v. FCC,* 512 U.S. 622 (1994). Similarly, the Supreme Court has rejected the notion that a listener has an absolute right "to listen only to such points of view as the listener wishes to hear." *Public Utilities Commission v. Pollak,* 343 U.S. 451, 463 (1952). As Lawrence Lessig says, "Government can . . . intervene to induce the codewriters to build into this code structures that preserve [democratic] values." Lessig, "The Law of the Horse."

23. *Kovacs v. Cooper,* 336 U.S. 77, 87 (1949) (emphasis added).

CHAPTER 20: SURF GLOBALLY, NETWORK LOCALLY

1. Charles Lockwood, "Rebuilding a Sense of Community," *Wall Street Journal,* August 29, 1997, A10.

2. See, for example, Ray Oldenberg, *The Great Good Place: Cafés, Coffee Shops, Community Centers, Beauty Parlors, General Stores, Bars, Hangouts and How They Get You through the Day* (New York: Marlowe and Company, 1989); George W. S. Trow, *Within the Context of No Context* (Boston: Little, Brown, and Co., 1981).

3. Quoted in Rheingold, *The Virtual Community,* 24.

4. On cable television, see Monroe E. Price and John Wicklein, *Cable Television: A Guide for Citizen Action* (Philadelphia: Pilgrim Press, 1972); Ralph Lee Smith, *The*

Wired Nation: Cable TV: The Electronic Communications Highway (New York: Harper and Row, 1972). On radio, see Susan Smulyan, *Selling Radio: The Commercialization of American Broadcasting, 1920–1934* (Washington: Smithsonian Institution Press, 1994); Robert W. McChesney, *Telecommunications, Mass Media, and Democracy: The Battle for the Control of U.S. Broadcasting, 1928–1935* (New York: Oxford University Press, 1993); Susan J. Douglas, *Inventing American Broadcasting, 1899–1922* (Baltimore: Johns Hopkins University Press, 1987); Todd Lappin, "Déjà Vu All Over Again," *Wired*, May 1995, 175.

5. See Gary Chapman and Lodis Rhodes, "Nurturing Neighborhood Nets," *Technology Review*, October 1997, 48, 51.

6. See www.nptn.org. Toward the end, the organization changed its name to the National Public Telecomputing Network.

7. A new coordinating organization, the Association for Community Networking, has also been established. See bcn.boulder.co.us/afcn.

8. See Peter Hinssen, "Life in the Digital City," *Wired*, June 1995, 90.

9. Quoted in Bruno Giusanni, "A Year and Half Later, a Wired Neighborhood Looks Back," *Cybertimes (New York Times)*, October 13, 1998, at www.nytimes.com/library/tech/98/10/cyber/eurobytes/13euro.html.

10. See Katie Hafner, "The World's Most Influential Online Community," *Wired*, May 1997, 98, 104.

11. Rheingold, *The Virtual Community*, 37.

12. Ibid., 209.

13. Ibid., 235.

14. See www.echonyc.com/about/more.html.

15. Studies routinely show that people are interested in news of their own communities. Nearly seven in ten respondents in one survey, for example, said they were interested in news from where they lived; about five in ten were interested in national news, and four in ten were interested in world news. Lawrie Mifflin, "Crime Falls, But Not on TV," *New York Times*, July 6, 1997, Week in Review, 4 (citing study by Roper Center for Public Opinion Research, the Newseum, and the Media Studies Center).

16. See Chapman and Rhodes, "Nurturing Neighborhood Nets," 48, 51. See also Stephen Doheny-Farina, *The Wired Neighborhood* (New Haven: Yale University Press, 1996).

17. Harvard law professor Charles Nesson speaks of the need for a "commons in cyberspace," which he uses in a slightly different but complementary manner to refer to public-oriented online resources.

18. In 1997, magazines like *Business Week* and books like *Net Gain,* by John Hagel III and Arthur A. Armstrong, were celebrating the marriage of commerce and community online. But a year later, the results of these corporate-backed communities were "markedly subpar," in the words of one trade journal, which noted that many elaborately constructed chat forums were drawing one user message a day or a few per month. Mark Gimein, "On Portal Row," *Industry Standard*, June 8, 1998, 28, 31.

19. Laurie J. Flynn, "The Real Internet Action is Local," *New York Times*, September 14, 1998, C9.

20. The signs so far are not that encouraging. Microsoft decided that Sidewalk, which

it launched as a group of community-based web sites, should not have open dialogue forums where users could communicate with one another and post their own thoughts. Apparently, the company feared they would not be able to control what was said there, especially if the forums attracted Microsoft critics. It therefore decided in late 1998 to recast Sidewalk not as a community site, but as a portal for commerce—in other words, yet another route to Microsoft's online stores. See Martin Wolk, "Microsoft Relaunching Sidewalk as Guide to Commerce," Reuters, October 21, 1998.

21. If a commercial online provider fails to make these commitments, Net users should vote with their feet and patronize one that does—or create one that does, perhaps in partnership with a local community organization, newspaper, or radio station. If necessary, government might also create incentives for local sites to play more of a role in community networking than they do now. Already, the U.S. government has recognized the value of community networking through grants from the National Telecommunications and Information Administration.

22. A cynic might note that this would be no different from how we live today. It is, of course, true that individual will and socioeconomic reality often combine to leave people segregated in their geographic communities. This unfortunate fact, though, strengthens the weight of my argument rather than lessening it.

CHAPTER 21: THE TOOLS OF DEMOCRACY

1. Esther Dyson, *Release 2.0*, 7.

2. The need for balance between markets and government, and an articulation of what each institution does well, can be found in Robert Kuttner, *Everything for Sale: The Virtues and Limits of Markets* (New York: Knopf, 1997), and in George Soros, *The Crisis of Global Capitalism: Open Society Endangered* (New York: PublicAffairs, 1998).

3. An example of such innovative governance was the emergence in 1998 of an international nonprofit organization, the Internet Corporation for Assigned Names and Numbers, to assume the administration of the Internet's domain name system. See www.icann.org.

4. Quoted in Cass R. Sunstein, "The Road from Serfdom," *New Republic*, October 20, 1997, 36, 40.

5. "Remarks by the President in Announcement of Electronic Commerce Initiative," White House Press Release, July 1, 1997.

6. See Children's Online Privacy Protection Act, Pub. L. No. 105-277, 105 Enacted H.R. 4328 (1998) (codified at various sections of title 13 of the United States Code).

7. The European Union's "Directive 95/46/EC of the European Parliament and of the Council of 24 October 1995 on the Protection of Individuals with Regard to the Processing of Personal Data and on the Free Movement of Such Data" went into effect on October 25, 1998. The directive and information about it are available at www.cdt.org/privacy/eudirective.

8. See John Markoff, "Differences Over Privacy on the Internet," *New York Times*, July 1, 1998, D1.

9. See, for example, the description of DigiCash at www.digicash.com/ecash/ecash-home.html.

10. In addressing such a possibility, Ira Magaziner, then President Clinton's chief Internet advisor, observed that "taxing authorities would save money by getting more compliance and getting money sooner." Matt Hamblen, Sharon Machlis, and Patrick Thibodeau, "Politicos Wrangle Over Internet Taxes; Partial Ban Could Derail Moratorium Bill," *ComputerWorld*, March 9, 1998, 1, 16.

11. For an interesting example of a digital intermediary that attempts to prevent fraud, see www.scambusters.com.

12. See 12 U.S.C. sec. 1828 (1998) (requiring banks insured by the Federal Deposit Insurance Corporation to post signs stating that deposits are federally insured to $100,000). In the online context, the ease of access to information—through metadata or linking—allows the consumer to learn much more about the vendor than simply whether it is government approved.

13. Examples of "public feedback regulation" include requirements that U.S. airlines publish on-time arrival records and statistics on lost baggage. When compelled to disclose this information, airlines obviously have a strong incentive to improve their performance so that they are closer to the top of the field than the bottom. Similar disclosure requirements have been used in the area of automobile safety (accident rates for each model), home mortgage lending (broken down by race, sex, and income), and environmental protection (list of toxic materials in various products). *See* Brin, *The Transparent Society,* 252–253.

14. Pool, *Technologies of Freedom,* 173.

15. Economist Garth Saloner and law professor David Post are among those who have argued that standards made dominant through network effects should perhaps lose their proprietary status. See David Post, "Opening Up Windows," *American Lawyer,* June 1998, 79.

16. As to the constitutionality of this idea, the eminent First Amendment scholar Zechariah Chafee said it well when he explained that speech markets, like markets for products, sometimes require regulation in order to work smoothly. As a result, he said, "government can lay down the rules of the game which will promote rather than restrict speech." Zechariah Chafee Jr., *Government and Mass Communications: A Report from the Commission on Freedom of the Press* (Chicago: University of Chicago Press, 1947), 471.

17. The U.S. Congress's Thomas database (thomas.loc.gov) actually contains 219 bills with the word "Internet" during the 105th Congress (1997–98). The same database shows twenty-five such bills during the 104th Congress (1995–96), and four during the 103rd Congress (1993–94). For a summary of the 105th Congress's most prominent Internet legislation, see Alan N. Sutin and Ellen Goldberg, "High-Tech Agenda in Congress," *New York Law Journal*, December 21, 1998, S3.

18. "Remarks by the President in Announcement of Electronic Commerce Initiative," White House Press Release, July 1, 1997.

19. See The National Information Infrastructure: Agenda for Action (1993), at www.usgs.gov/public/nii/NII-Executive-Summary.html. The 1998 report, "U.S. Government Working Group On Electronic Commerce, First Annual Report," can be found, along with other Internet policy statements, at www.ecommerce.gov.

20. See, for example, Simson L. Garfinkel, "The Web's Unelected Government," *Technology Review*, November/December 1998, 38.

21. Davidson and Rees-Mogg, *The Sovereign Individual,* 320.

22. Jean Bethke Elshtain, "Authority Figures," *New Republic*, December 22, 1997, 11–12.

23. Handy, *The Hungry Spirit*, 63.

EPILOGUE: FROM REVOLUTION TO RESOLUTION

1. Learned Hand, *The Spirit of Liberty: Papers and Addresses of Learned Hand*, 2d ed.(New York: Knopf, 1953), 189–190. See also Gerald Gunther, *Learned Hand: The Man and the Judge* (New York: Knopf, 1994), 547–552 (discussing the speech).

2. Quoted in Cheryl Lavin, "Fast Track; Replays," *Chicago Tribune Sunday Magazine*, January 7, 1996, 8.

Bibliography

Abelson, Robert P., et al., eds. *Theories of Cognitive Consistency: A Sourcebook.* Chicago: Rand McNally and Co., 1968.

Abramson, Jeffrey B., F. Christopher Arterton, and Gary R. Orren. *The Electronic Commonwealth: The Impact of New Media Technologies on Democratic Politics.* New York: Basic Books, 1988.

Ackerman, Bruce. *Social Justice in the Liberal State.* New Haven: Yale University Press, 1980.

Agre, Philip E., and Marc Rotenberg, eds. *Technology and Privacy: The New Landscape.* Cambridge, Mass.: MIT Press, 1997.

Anderson, Benedict. *Imagined Communities: Reflections on the Origin and Spread of Nationalism.* London: Verso, 1983.

Arterton, F. Christopher. *Teledemocracy: Can Technology Protect Democracy?* Newbury Park, Calif.: Sage Publications, 1987.

Bagdikian, Ben H. *Media Monopoly.* 5th ed. Boston: Beacon Press, 1997.

Barber, Benjamin. *Strong Democracy: Participatory Politics for a New Age.* Berkeley: University of California Press, 1984.

Bellah, Robert N., et al. *Habits of the Heart: Individualism and Commitment in American Life.* Berkeley: University of California Press, 1985.

Beniger, James R. *The Control Revolution: Technological and Economic Origins of the Information Society.* Cambridge, Mass.: Harvard University Press, 1986.

Berlin, Isaiah. *Four Essays on Liberty.* New York: Oxford University Press, 1969.

Birkerts, Sven. *The Gutenberg Elegies: The Fate of Reading in an Electronic Age.* New York: Fawcett Columbine, 1994.

Bittner, John R. *Law and Regulation of Electronic Media.* 2d ed. Englewood Cliffs, N.J.: Prentice-Hall 1994.

Boyle, James. *Shamans, Software, and Spleens: Law and the Construction of the Information Society.* Cambridge, Mass.: Harvard University Press, 1996.

Brand, Steward. *The Media Lab: Inventing the Future at MIT.* New York: Viking, 1987.

Branscomb, Anne W. *Toward a Law of Global Communications Networks.* New York: Longman, 1986.

————. *Who Owns Information? From Privacy to Public Access.* New York: Basic Books, 1994.

Briggs, Asa. *The BBC: The First Fifty Years.* New York: Oxford University Press, 1985.

Brin, David. *The Transparent Society: Will Technology Force Us to Choose Between Privacy and Freedom?* Reading, Mass.: Addison-Wesley, 1998.

Brook, James, and Iain A. Boal, eds. *Resisting the Virtual Life: The Culture and Politics of Information.* San Francisco: City Lights Books, 1995.

Browning, Graeme. *Electronic Democracy: Using the Internet to Affect American Politics.* Wilton, Conn.: Pemberton Press Books/Online Inc., 1996.

Burke, Edmund. *Reflections on the Revolution in France*. Vol. 1. London: John Sharpe, 1820.

Burstein, Daniel, and David Kline. *Road Warriors: Dreams and Nightmares Along the Information Highway*. New York: Plume, 1996.

Carey, James W. *Communication as Culture: Essays on Media and Society*. New York: Routledge, 1992.

Caristi, Dom. *Expanding Free Expression in the Marketplace: Broadcasting and the Public Forum*. New York: Quorum Books, 1992.

Chafee, Zechariah. *Government and Mass Communication: A Report from the Commission on Freedom of the Press*. Chicago: University of Chicago Press, 1947.

Collins, Ronald K. L., and David M. Skover. *The Death of Discourse*. Boulder: Westview Press, 1996.

Cronin, Thomas E. *Direct Democracy: The Politics of Initiative, Referendum, and Recall*. Cambridge, Mass.: Harvard University Press, 1989.

Crowley, David, and Paul Heyer, eds. *Communication in History*. 2d ed. New York: Longman, 1995.

Dahl, Robert. *Democracy and Its Critics*. New Haven: Yale University Press, 1989.

Davidson, James Dale, and Lord William Rees-Mogg. *The Sovereign Individual: How To Survive and Thrive During the Collapse of the Welfare State*. New York: Simon and Schuster, 1997.

Diffie, Whitfield, and Susan Landau. *Privacy on the Line: The Politics of Wiretapping and Encryption*. Cambridge, Mass.: MIT Press, 1998.

Doheny-Farina, Stephen. *The Wired Neighborhood*. New Haven: Yale University Press, 1996.

Douglas, Susan J. *Inventing American Broadcasting, 1899–1922*. Baltimore: Johns Hopkins University, 1987.

Drake, William J., ed. *The New Information Infrastructure: Strategies for U.S. Policy*. New York: Twentieth Century Fund Press, 1995.

Dyson, Esther. *Release 2.0: A Design for Living in the Digital Age*. New York: Broadway Books, 1997.

Eisenstein, Elizabeth L. *The Printing Revolution in Early Modern Europe*. New York: Cambridge University Press, 1983.

Ellul, Jacques. *The Technological Society*. New York: Knopf, 1964.

The Emerging Internet. Queenstown, Md.: Aspen Institute, Institute for Information Studies, 1998.

Ermann, M. David, Mary B. Williams, and Michele S. Shauf, eds. *Computers, Ethics, and Society*. 2d ed. New York: Oxford University Press, 1997.

Etzioni, Amitai, ed. *New Communitarian Thinking: Persons, Virtues, Institutions, and Communities*. Charlottesville: University Press of Virginia, 1995.

Festinger, Leon. *Conflict, Decision, and Dissonance*. Stanford: Stanford University Press, 1964.

———. *A Theory of Cognitive Dissonance*. Evanston, Ill.: Row, Peterson, 1957.

Fishkin, James S. *Democracy and Deliberation: New Directions for Democratic Reform*. New Haven: Yale University Press, 1991.

———. *The Voice of the People: Public Opinion and Democracy*. New Haven: Yale University Press, 1995.

Fiss, Owen M. *The Irony of Free Speech*. Cambridge, Mass.: Harvard University Press, 1996.

————. *Liberalism Divided: Freedom of Speech and the Many Uses of State Power.* Boulder: Westview Press, 1996.

Flaherty, David. *Protecting Privacy in Surveillance Societies.* Chapel Hill, N.C.: University of North Carolina Press, 1989.

Flamm, Kenneth. *Creating the Computer: Government, Industry, and High Technology.* Washington, D.C.: Brookings Institution, 1988.

Forrester, Tom, and Perry Morrison. *Computer Ethics: Cautionary Tales and Ethical Dilemmas in Computing.* Cambridge, Mass.: MIT Press, 1990.

Foucault, Michel. *Discipline and Punish: The Birth of the Prison.* Edited and translated by Alan Sheridan. New York: Pantheon Books, 1979.

Friendly, Fred W. *The Good Guys, The Bad Guys and the First Amendment: Free Speech vs. Fairness in Broadcasting.* New York: Random House, 1975.

Frank, Thomas. *The Conquest of Cool: Business Culture, Counterculture, and the Rise of Hip Consumerism.* Chicago: University of Chicago Press, 1997.

Fromm, Erich. *Escape from Freedom.* New York: Rinehart and Company, 1941.

Fuller, Buckminster. *No More Secondhand God, and Other Writings.* Carbondale, Ill.: Southern Illinois University Press, 1963.

Gallup, George. *A Guide to Public Opinion Polls.* Princeton, N.J.: Princeton University Press, 1948.

Gandy, Oscar H. *The Panoptic Sort: A Political Economy of Personal Information.* Boulder: Westview Press, 1993.

Gans, Herbert J. *Middle American Individualism: The Future of Liberal Democracy.* New York: Oxford University Press, 1988.

Gates, William H. *The Road Ahead.* Rev. ed. New York: Penguin, 1996.

Gibson, William. *Neuromancer.* New York: Ace Books, 1984.

Gilder, George. *Life After Television.* Rev. ed. New York: W. W. Norton, 1994.

Gitlin, Todd. *The Twilight of Common Dreams: Why America Is Wracked By Culture Wars.* New York: Metropolitan Books, 1995.

Gleick, James. *Chaos: Making a New Science.* New York: Penguin Books, 1987.

Godwin, Mike. *Cyber Rights: Defending Free Speech in the Digital Age.* New York: Times Books, 1998.

Graber, Mark A. *Transforming Free Speech: The Ambiguous Legacy of Civil Libertarianism.* Berkeley: University of California, 1991.

Grossman, Lawrence K. *The Electronic Republic: Reshaping Democracy in the Information Age.* New York: Twentieth Century Fund/Viking Press, 1995.

Grossman, Wendy M. *Net.wars.* New York: New York University Press, 1997.

Gurak, Laura, J. *Persuasion and Privacy in Cyberspace: The Online Protests over Lotus Marketplace and the Clipper Chip.* New Haven: Yale University Press, 1997.

Habermas, Jurgen. *Between Facts and Norms.* Cambridge, Mass.: MIT Press, 1996.

Hafner, Katie, and Matthew Lyons. *Where Wizards Stay Up Late: The Origins of the Internet.* New York: Simon and Schuster, 1996.

Hagel, John, and Arthur G. Armstrong. *Net Gain: Expanding Markets Through Virtual Communities.* Boston: Harvard Business School Press, 1997.

Hand, Learned. *The Spirit of Liberty: Papers and Addresses of Learned Hand.* 2d ed. New York: Knopf, 1953.

Handy, Charles. *The Hungry Spirit: Beyond Capitalism: A Quest for Purpose in the Modern World.* New York: Broadway Books, 1998.

Hixson, Richard F. *Privacy in a Public Society: Human Rights in Conflict.* New York: Oxford University Press, 1987.

Horwitz, Robert Britt. *The Irony of Regulatory Reform.* New York: Oxford University Press, 1989.

Innis, Harold A. *The Bias of Communication.* Buffalo, N.Y.: University of Toronto Press, 1951.

Johnson, Steven. *Interface Culture: How New Technology Transforms the Way We Create and Communicate.* San Francisco: HarperEdge, 1997.

Kahin, Brian, and James Keller, eds. *Public Access to the Internet.* Cambridge, Mass.: MIT Press, 1995.

Kahin, Brian, and Charles Nesson, eds. *Borders in Cyberspace: Information Policy and Global Information Infrastructure.* Cambridge, Mass.: MIT Press, 1997.

Katsh, M. Ethan. *The Electronic Media and the Transformation of Law.* New York: Oxford University Press, 1989.

———. *Law in a Digital World.* New York: Oxford University Press, 1995.

Kelly, Kevin. *Out of Control: The Rise of Neo-biological Civilization.* Reading, Mass.: Addison-Wesley, 1994.

Kuttner, Robert. *Everything for Sale: The Virtues and Limits of Markets.* New York: Knopf, 1997.

Lasch, Christopher. *The Revolt of the Elites and the Betrayal of Democracy.* New York: W. W. Norton, 1996.

Lefcourt, Herbert M. *Locus of Control: Current Trends in Theory and Research.* 2d ed. Hillsdale, N.J.: Lawrence Erlbaum Associates, 1982.

Lerner, Daniel. *The Passing of Traditional Society.* Glencoe, Ill.: Free Press, 1958.

Levinson, Paul. *The Soft Edge: A Natural History of the Information Revolution.* New York: Routledge, 1997.

Levy, Steven. *Hackers: Heroes of the Computer Revolution.* Garden City, N.Y.: Anchor Press/Doubleday, 1984.

Liebling, A. J. *The Press.* New York: Ballantine Books, 1975.

Lippmann, Walter. *The Public Philosophy.* New York: New American Library, 1955.

Loader, Brian D., ed. *The Governance of Cyberspace.* New York: Routledge, 1997.

Ludlow, Peter, ed. *High Noon on the Electronic Frontier: Conceptual Issues In Cyberspace.* Cambridge, Mass.: MIT Press, 1996.

Mansell, Robin, and Roger Silverstone. *Communication By Design: The Politics of Information and Communication Technologies.* New York: Oxford University Press, 1996.

Marx, Karl. *The Eighteenth Brumaire of Louis Bonaparte.* New York: International Publishers, 1963.

McChesney, Robert W. *Telecommunications, Mass Media, and Democracy: The Battle for Control of U.S. Broadcasting, 1928–1935.* New York: Oxford University Press, 1993.

McChesney, Robert W., and Edward S. Herman. *The Global Media: The New Missionaries of Corporate Capitalism.* Washington, D.C.: Cassell, 1997.

McLuhan, Marshall. *Understanding Media: The Extensions of Man.* New York: McGraw-Hill, 1964.

Mead, George H. *Mind, Self and Society.* Chicago: University of Chicago, 1934.

Meyrowitz, Joshua. *No Sense of Place: The Impact of Electronic Media on Social Behavior.* New York: Oxford University Press, 1985.

Miller, Arthur M. *The Assault on Privacy: Computers, Data Banks, and Dossiers.* Ann Arbor, Mich.: University of Michigan Press, 1971.

Miller, Steven S. *Civilizing Cyberspace: Policy, Power, and the Information Superhighway.* Reading, Mass.: Addison-Wesley, 1996.

Mills, Stephanie, ed. *Turning Away from Technology: A New Vision for the 21st Century.* San Francisco: Sierra Club Books, 1997.

Mitchell, William J. *City of Bits: Space, Place, and the Infobahn.* Cambridge, Mass.: MIT Press, 1995.

Mosco, Vincent. *The Pay-Per Society: Computers and Communication in the Information Age.* Toronto: Garamond Press, 1989.

Mosco, Vincent, and Janet Wasko. *The Political Economy of Information.* Madison, Wisc.: University of Wisconsin Press, 1988.

Mumford, Lewis. *Technics and Human Development: The Myth of the Machine.* New York: Harcourt Brace Jovanovich, 1967.

Naisbitt, John. *Megatrends.* New York: Warner Books, 1982.

Negroponte, Nicholas. *Being Digital.* New York: Knopf, 1995.

Neuman, W. Russell. *The Future of the Mass Audience.* New York: Cambridge University Press, 1991.

Norman, Donald A. *The Invisible Computer: Why Good Products Can Fail, the Personal Computer Is So Complex, and Information Appliances Are the Solution.* Cambridge, Mass.: MIT Press, 1998.

Nye, Joseph S., Jr., Philip D. Zelikow, and David C. King, eds. *Why People Don't Trust Government.* Cambridge, Mass.: Harvard University Press, 1997.

Oldenberg, Ray. *The Great Good Place: Cafés, Coffee Shops, Community Centers, Beauty Parlors, General Stores, Bars, Hangouts and How They Get You Through the Day.* New York: Marlowe and Company, 1989.

Perlmutter, Lawrence C., and Richard Monty. *Choice and Perceived Control.* Hillsdale, N.J.: Lawrence Erlbaum Associates, 1979.

Pool, Ithiel de Sola, ed. *Talking Back: Citizen Feedback and Cable Technology.* Cambridge, Mass.: MIT Press, 1973.

———. *Technologies of Freedom.* Cambridge, Mass.: Belknap Press, 1983.

———. *Technologies Without Boundaries.* Cambridge, Mass.: Harvard University Press, 1990.

Porter, David, ed. *Internet Culture.* New York: Routledge, 1997.

Postman, Neil. *Technopoly: The Surrender of Culture to Technology.* New York: Knopf, 1992.

Price, Monroe E., and John Wicklein. *Cable Television: A Guide for Citizen Action.* Philadelphia: Pilgrim Press, 1972.

Renshon, Stanley Allen. *Psychological Needs and Political Behavior: A Theory of Personality and Political Efficacy.* New York: The Free Press, 1974.

Rheingold, Howard. *The Virtual Community: Homesteading on the Electronic Frontier.* Reading, Mass.: Addison-Wesley, 1993.

Riesman, David. *The Lonely Crowd: A Study of the Changing American Character.* New Haven: Yale University Press, 1950.

Rosenkranz, E. Joshua. *Buckley Stops Here: Loosening the Judicial Stranglehold on Campaign Finance Reform.* New York: Twentieth Century Fund, 1998.

Roszak, Theodore. *The Cult of Information: A Neo-Luddite Treatise on High Tech, Ar-*

tificial Intelligence, and the True Art of Thinking. Berkeley: University of California Press, 1994.

Rushkoff, Douglas. *Cyberia: Life in the Trenches of Hyperspace.* Harper San Francisco, 1994.

———. *Playing the Future: How Kids' Culture Can Teach Us to Thrive in an Age of Chaos.* New York: HarperCollins, 1996.

Saari, David J. *Too Much Liberty?: Perspectives on Freedom and the American Dream.* Westport, Conn.: Praeger, 1995.

Sardar, Ziauddin, and Jerome R. Ravetz, eds. *Cyberfutures: Culture and Politics on the Information Superhighway.* New York: New York University Press, 1996.

Savan, Leslie. *The Sponsored Life: Ads, TV, and American Culture.* Philadelphia: Temple University Press, 1994.

Schlesinger, Arthur M. *The Disuniting of America.* New York: W. W. Norton, 1992.

Schmookler, Andrew Bard. *The Illusion of Choice: How the Market Economy Shapes Our Destiny.* Albany, N.Y.: SUNY Press, 1993.

Schuler, Douglas. *New Community Networks: Wired for Change.* Reading, Mass.: Addison-Wesley, 1996.

Sclove, Richard E. *Democracy and Technology.* New York: Guilford Press, 1995.

Segal, Howard P. *Future Imperfect: The Mixed Blessings of Technology in America.* Amherst, Mass.: University of Massachusetts Press, 1994.

Shenk, David. *Data Smog: Surviving the Information Glut.* New York: HarperCollins, 1997.

Shiffrin, Steven H. *The First Amendment, Democracy, and Romance.* Cambridge, Mass.: Harvard University Press, 1990.

Slavko, Splichal, Andrew Calabrese, and Colin Sparks, eds. *Information Society and Civil Society.* West Lafayette, Ind.: Purdue University Press, 1994.

Slouka, Mark. *War of the Worlds: Cyberspace and the High-Tech Assault on Reality.* New York: Basic Books, 1995.

Smith, Ralph Lee. *The Wired Nation: Cable TV: The Electronic Communications Highway.* New York: Harper and Row, 1972.

Smulyan, Susan. *Selling Radio: The Commercialization of American Broadcasting, 1920–1934.* Washington: Smithsonian Institution Press, 1994.

Soros, George. *The Crisis of Global Capitalism: Open Society Endangered.* New York: PublicAffairs, 1998.

Stoll, Clifford. *Silicon Snake Oil: Second Thoughts on the Information Highway.* New York: Doubleday, 1995.

Sunstein, Cass R. *Democracy and the Problem of Free Speech.* New York: Free Press, 1993.

Talbott, Stephen L. *The Future Does Not Compute: Transcending the Machines in Our Midst.* Sebapastopol, Calif.: O'Reilly and Associates, 1995.

Tehranian, Majid. *Technologies of Power: Information Machines and Democratic Prospects.* Norwood, N.J.: Ablex Publishing Company, 1990.

Tenner, Edward. *Why Things Bite Back: Technology and the Revenge of Unintended Consequences.* New York: Knopf, 1996.

Toffler, Alvin. *Powershift: Knowledge, Wealth, and Violence in the 21st Century.* New York: Bantam Books, 1990.

Trow, George W. S. *Within the Context of No Context.* Boston: Little, Brown, and Co., 1981.

Turkle, Sherry. *Life on the Screen: Identity in the Age of the Internet.* New York: Simon and Schuster, 1995.

Turow, Joseph. *Breaking Up America: Advertisers and the New Media World.* Chicago: University of Chicago Press, 1997.

Ullman, Ellen. *Close to the Machine: Technophilia and Its Discontents.* San Francisco: City Lights Books, 1997.

Viorst, Judith. *Imperfect Control: Our Lifelong Struggle with Power and Surrender.* New York: Simon and Schuster, 1998.

Westin, Alan. *Privacy and Freedom.* New York: Atheneum, 1967.

Wiener, Norbert. *Cybernetics: Or Control and Communication in the Animal and the Machine.* Cambridge, Mass.: MIT Press, 1961.

———. *The Human Use of Human Beings: Cybernetics and Society.* Boston: Houghton Mifflin, 1954.

Winner, Langdon. *Autonomous Technology: Technics-Out-of-Control as a Theme in Political Thought.* Cambridge, Mass: MIT Press, 1977.

———. *The Whale and the Reactor: A Search for Limits in an Age of High Technology.* Chicago: University of Chicago Press, 1986.

Woodward, Kathleen, ed. *The Myths of Information: Technology and Postindustrial Culture.* Madison, Wisc.: Coda Press, 1980.

Wresch, William. *Disconnected: Haves and Have-nots in the Information Age.* New Brunswick, N.J.: Rutgers University Press, 1996.

Wriston, Walter B. *The Twilight of Sovereignty: How the Information Revolution Is Transforming Our World.* New York: Charles Scribner's Sons, 1992.

Acknowledgments

▶ I've lived the central theme of this book while writing it. Even while luxuriating in the autonomy that new technology makes possible, I have recognized my own need for balance—particularly in terms of reliance on friends, family, colleagues, mentors, and institutions.

My first debt is to my teachers, who have been not just inspiring sages but true friends. I first became interested in emerging communications technology because of my studies at Yale Law School with Professors Bruce Ackerman, Owen Fiss, and Lawrence Lessig. I can't thank them enough for pushing me to think and write—and for listening to my ideas along the way. This book has been greatly influenced by their work, particularly Lessig's groundbreaking writings about the Internet and the law.

Many thanks to Richard C. Leone, Richard Ravitch, Greg Anrig, and everyone else associated with The Century Foundation for generously supporting this project and for finding other ways for us to collaborate while I was in residence there from October 1996 through 1997.

I was honored when Peter Osnos decided that he wanted this book to be among the first that he signed for his distinguished new publishing house, PublicAffairs. Peter's advice has guided me throughout this project, as has that of his patient and good-humored managing editor, Robert Kimzey.

I am immensely grateful to those who have given me a nurturing environment in which to work and write: Charles Nesson, Lawrence Lessig, Jonathan Zittrain, and the scores of folks linked to the Berkman Center for Internet and Society at Harvard Law School, where I spent the first half of 1998; Burt Neuborne, Joshua Rosenkranz, and the outstanding lawyers and staff at the Brennan Center for Justice at NYU Law School, where I have been fortunate to hang my hat since Labor Day 1998; and Charles Firestone and the rest of the stellar team at the Aspen Institute, a unique organization for which I am proud to work.

Though I have been lucky to write for a number of different publications, *The Nation* has been my writing home since I was an intern there in 1991, and I thank Victor Navasky, Katrina vanden Heuvel, and the others at the magazine for their ongoing support. A few parts of the book—particularly Chapter 15—benefited from dress rehearsal in the pages of *The Nation*, and I pursued this project in large part because of the response to "Street Corners in Cyberspace," a 1995 piece I wrote for the magazine about free speech and the commercialization of the Internet. (Thanks also to Project Censored, which named it one of its top stories of the year.)

During the course of this project, I was fortunate to reestablish ties with David Shenk and Steven Johnson, classmates of mine from Brown (what was in the water up there?) whose insightful writings about new technology have taught me a tremendous amount. In spring of 1998, along with nine other writers, we launched the technorealism project (www.technorealism.org). Thanks to the other original technorealists for joining in the fun and to the thousands who lent their support—foremost among them Mitchell Kapor, whose prescient early work in this area embodied the technorealist approach before it had a name.

Special gratitude goes to those who read manuscript drafts or otherwise pushed the book along at critical stages: Peter Blake, Alan Davidson, Steve Gunn, Pierre N. Leval, Aaron Naparstek, Abigail Phillips, Josh Rosenkranz, Jon Rubin, Daniel Shapiro, David Shenk, and Olivier Sultan. (A few of you went beyond the call of duty and I am forever indebted.) I also received first-rate research assistance from Melanie Glickson, April Harshaw, Jill Steinberg, and Federico Windhausen.

Over the past few years, I've benefited from conversations about the Net (some virtual, some heated, some both) with too many people to mention here, but you know who you are and I tip my hat to you.

During a sometimes arduous writing process, I have also been sustained in myriad ways by my remarkable parents, brothers, and extended family, and by a circle of loving friends who have somehow put up with me. But no one deserves my thanks and praise more than Nina Bauer, who tolerated my often absurd work schedule with a sunny outlook and unflagging support. This is for you, love. And yes, it's really done.

Index

Abramson, Jeffrey, 113–114
Accessibility, 17–18, 42, 180, 182, 224–227
Activism, 37–38, 49–52, 195, 214–215. *See also* Advocacy groups
Adler, Alfred, 22
Advanced Research Projects Agency, 20
Advocacy groups, 228–229. *See also* Activism
Algeria, 42
Alphabetic writing, 32–33
Amazon.com, 55–56, 98, 142, 146, 222
America Online (AOL), 44, 91, 97–98, 100, 131, 180, 181, 183, 204, 214
American Enterprise Institute, 134
Amsterdam, 211
Analog media, 16
Anderson, Benedict, 116
Anti-Milosevic protesters, 7, 9
Antisocial behavior, 119–120
Antitrust battles, 87–88, 89–92, 182–184
AOL. *See* America Online
Apple computers, 26–27, 185
Arpanet, 20
"Avatars," 52

Bahrain, 67
Barlow, John Perry, 30–31, 112
Batson, Bill, 132
Bavaria, 68
BBSs. *See* Computer bulletin-board systems
Belgium, 22
Belgrade, 7
Bereano, Philip, 163
Berlin, Isaiah, 22
Berners-Lee, Tim, 199
Bettmann Archive, 87
Biotechnology, 11–12

Blacksburg Electronic Village, 210, 214, 227
Blumenthal, Sidney, 134
Bolt Reporter, 42
Bork, Robert, 164
Boston Globe, 141
Brand, Stewart, 105–106
Brill's Content, 190
Broadband networks, 17
Broadcast regulations, 19–20, 169
Browsers
software, 90–91, 173–174, 182–183
"kid", 173–174
Bundled software, 182
BurmaNet, 50
Bush, George, 3
Bylines (web site), 56–57

Canada, 22
Cardozo, Benjamin, 148
Carlin, George, 71–72
Carnegie Mellon University, 34–35, 37
CDA. *See* Communications Decency Act
Censorship, 42, 51. *See also* Pornography; Speech
Center for Democracy and Technology, 51
Central America, 6
Channel capacity, 19
Chaos, 197–207
Chat rooms, 49, 108
Children
and browsers, 173–174
and sexual content, 170–174
and technology, 202–203
China, 6, 7, 42, 65–66
Choices on the Internet, 180–186
Christian Coalition, 115, 118

CIX, 212
"Clickwrap contracts," 80–81, 178, 179
Clinton, Bill, 19–20, 41, 48, 70, 79, 135, 158, 204, 220–221
Clipper Chip, 76
CNN, 141
COARA, 212
Code of the Internet, 14, 15–18, 37, 44, 56, 108, 169, 180–181, 205, 222–224, 228
Code regulation, 62, 69, 70, 72–73, 76–78, 80, 82, 83, 108, 177, 179, 222–224, 228–229
Cognitive dissonance, 109–111
Cold War, 6
Columbia Journalism Review, 190
Commercialization, 214, 222–224
Common Sense (Paine), 5
Communications Decency Act (CDA), 19, 38, 51, 70–73, 115, 171–172, 203–204
Communications
  and government, 224–227. *See also* Federal government
  industry. *See* Mass media
  remote, 193–194
  resources, 19, 224
  revolution, 9, 10
  and technology, 12, 14. *See also* Internet; Online communities
Communities, 116–120, 208–216. *See also* Online communities
Computer bulletin board systems (BBSs), 35
Computers
  and control, 25–33
  IBM. *See* IBM
  Macintosh. *See* Macintosh computers
  network, 92
  personal (PCs), 26–27, 88, 93–95
  See also Technology
Confrontation, 126–129
Consumer Project on Technology (CPT), 93
Consumerism, 56, 142–149, 222–224
Control revolution, 10–11. *See also* Corporate control; Federal government,

control; Individualism; Internet; Personalization; Randomness
Control
  and computers, 25–33
  corporate, 84–100
  of creativity, 78–83
  government. *See* Federal government, control
  psychology of, 22–24
  revolution, 21, 22, 62
  of secrets, 73–78
  of speech, 64–73, 108, 124–132, 203–207
Copyright, 78–82, 170, 177–179
Coray, Richard, 71, 72
Corporate control, 84–100
CPT. *See* Consumer Project on Technology
Creativity, 42–43, 78–82, 177–179
Crime, 150–153
Cuba, 42
Cybernetics, 32
Cyberporn. *See* Pornography
Cyber-rights, 51
Cyberspace, 29–33

Dallas Morning News, 138
Dalzell, Stewart, 203–204
Data Smog (Shenk), 44–45
Davidson, James Dale, 229
Davis, Richard Allen, 150
Day trading, 54–55, 145–146
Decentralization, 57–59
"Declaration of the Independence of Cyberspace" (Barlow), 30
Defense Department, 20
Delegation, 187–196
Deliberative polling, 194
Democracy, 15, 153–157, 192–196, 217–230
Deregulation, 19–20, 217–218
Digital activism, 37–38
Digital certificates, 173
Digital Freedom Network, 42
Digital information, 16
Digital revolution, 9
Digital watermarks, 80

Dionne, E. J., 120
Direct democracy. *See* Democracy
Direct Marketing Association, 221
Disintermediation, 55–57, 133–141
  and consumerism, 142–149
  and journalism, 133–141, 188–192
  and politics, 150–158, 192–196
  and privacy, 158–165
  See also Intermediaries
Displacement, 111–112
Dissonance, 109–111
Distance learning, 202–203
Distributed networks, 16
Domain names, 20
Douglass, Frederick, 132
Drudge, Matt, 41, 133–141, 188
Durant, Will, 201
Dyson, Esther, 121, 218

East Timor, 50
Eastern Europe, 5, 6, 7
E-Bay, 57
E-cash, 222
Echo (online service), 212
Eco, Umberto, 139
Education, 57, 201–203
EFF. *See* Electronic Frontier Foundation
Egghead Software, 143
Electronic commerce. *See* Consumerism
Electronic Frontier Foundation (EFF),
  35, 36
Electronic mail, 42, 49, 108
Ellul, Jacques, 199–200
Elmer-DeWitt, Philip, 34–37, 55
Elshtain, Jean Bethke, 230
E-mail. *See* Electronic mail
Employment, 48, 143–145
Encryption, 73–78, 170, 175–177
E-rate plan, 224
Ethics of non-power, 199–200
Ethnicity, 51
"Existing rules" approach, 170, 175,178
European Union, 164, 221
Exon, James, 35

"Fair use," 79, 82, 178–179
Fairness and Accuracy in Reporting, 190

Fallows, James, 191
Faxes, 3–4, 6, 8
FBI. *See* Federal Bureau of Investigation
Federal Bureau of Investigation (FBI),
  75–76
Federal government, 19–20
  antitrust lawsuits, 89–91
  control, 63–83, 217, 220–221
  decentralization, 58
  and Internet, 19–20, 69–83, 227–230
  and open-source software, 185
  trust in, 21–22
Federal Trade Commission (FTC), 195,
  220–221
Feedback effect, 113–114
Festinger, Leon, 109, 110–111
Filtering, 45–46, 107–109, 125–126,
  172–173, 199. *See also* News per-
  sonalization services
Finney, Hal, 75
First Amendment. *See* Speech
Fishkin, James, 194
Fiss, Owen, 127
Franklin, Jon, 56–57
"Free-agent nation," 48
Freedom of speech. *See* Speech
Freeh, Louis, 75, 76
Free market, 18–21, 129–132, 217–218. *See
  also* Antitrust battles; Con-
  sumerism
Free-Nets, 210
Freeware. *See* Open-source software
Freud, Sigmund, 109–110
Frisch, Max, 233
Fromm, Erich, 186
FTC. *See* Federal Trade Commission
Fuller, Buckminster, 58

Gandy, Oscar, 162
Gatekeepers. *See* Accessibility
Gates, Bill, 25, 28, 84, 86, 129, 130, 142
Gateways, 212–215
Germany, 68–70
Gibson, William, 32
Ginsberg, Sam, 71, 72, 171, 174
Godwin, Mike, 35, 36, 38
Gorbachev, Mikhail, 3

Gore, Al, 160–161
Government control. *See* Federal government, control
The Graduate, 44
Graphical interfaces, 27
Grateful Dead, 30
Great Britain, 22
Gun control, 13–14
Gup, Ted, 197–198
Gutenberg, 5

Hackers (Levy), 25
Hand, Learned, 231–232
Handguns, 13–14
Handy, Charles, 230
Harvard University, 133
Hayek, Friedrich, 219
Hibbitts, Bernard J., 191
History
    of communities, 116–117
    Internet, 20, 209
    printing, 5–6
Hoffman, Donna, 36, 37
Hoffman, Dustin, 44
Holmes, Oliver Wendell, 129
HomeNet study, 122
Home schooling, 202–203
Htun, Aung Gyaw, 60, 66

IBM, 27, 89–90, 93
"Imagined communities," 116
Impact of technology, 13–15
Indecency-blocking software, 172
Individual empowerment, 11–12, 21–22
Individualism, 21–22, 65. *See also* Communities; Corporate control
Indonesia, 66
Infobeat, 45
Information
    digital, 16
    feedback loops, 113–114
    online. *See* Online information
    personal. *See* Privacy
    revolution, 9, 10
    secret, 73–78
Institutional control. *See* Corporate control; Federal government, control

Interactivity, 15, 34–43, 45, 137
Intermediaries, 172–173, 187–196, 222–224. *See also* Disintermediation
Internet
    accessibility. *See* Accessibility
    choices, 180–186
    code, 14, 15–18, 37, 44, 56, 108, 169, 180–181, 205, 222–224, 228
    and convenience, 180–186
    definition, 14–15
    and democracy, 15, 153–157, 192–196, 217–230
    and federal government, 19–20. *See also* Federal government
    and Freedom of Speech. *See* Speech
    international use of, 7–8, 65–67
    legal issues, 148–149, 169–179
    and pornography, 34–35
    privatization of, 20–21
    protocols, 184
    rules and contexts, 169–179
Internet Explorer (Microsoft), 90–91, 95, 182–183
Internet Relay Chat (IRC) channels, 42
Interoperability, 16–17
Investing. *See* Online investing
Iran, 67
IRC. *See* Internet Relay Chat channels
Ireland, 22
Italy, 22

Japan, 212
Java (Sun Microsystems), 91–92, 183
The Jerusalem Post, 47
Johnson, Steven, 27
Jordan, 67
Jordan, Vernon, 135
Journalism, 38–39, 133–141, 188–192. *See also* Mass media; News personalization services
Justice Department, 91, 183–184, 226–227

Katz, Jon, 38–39
Kelly, Kevin, 188
Key escrow, 76–77, 175–177

Kid browsers, 173–174
Kinsley, Michael, 86
Klaas, Marc, 150
Kristol, William, 135
Kurosawa, Akira, 189
Kurtz, Howard, 39
Kuwait, 67

Laissez-faire economics, 18–19, 20, 100,
    217–218
Language, 32–33, 51
Law enforcement, 75–78, 175–177
Leahy, Patrick, 51
Legal issues, 148–149, 169–179. *See also*
    Antitrust battles, Censorship,
    Copyright, Privacy, Speech
Lerner, Daniel, 122
Lessig, Lawrence, 15
Levin, Richard, 201–202
Levitt, Arthur, 54, 143
Levy, Steven, 25
Lewinsky, Monica, 41, 135
Liberal education. *See* Education
Libraries, 171
Licklider, J. C. R., 209
Liebling, A. J., 40, 41, 182
Lin, Hai, 65–66
Linux operating system, 184, 185
Lippmann, Walter, 153
Localism. *See* Communities
Locke, John, 229
Lockwood, Charles, 208
Loka Institute, 143
London, 211, 212
Luther, Martin, 5

Macintosh computers, 27, 28, 33, 185
Malaysia, 66
Many-to-many interactivity, 15, 137
"Market for privacy," 159–165, 220
Marketplace. *See* Free market
Marquardt, Angela, 69
Marzorati, Gerald, 200
Mass customization, 47–48
Mass media, 38–39, 117–118, 225
    analog, 16
    history, 6, 14

ownership, 19, 40–41
    See also Journalism; Television
McCurry, Mike, 141
Meeks, Brock, 36
Microsoft, 27, 44, 84–97, 173, 181,
    182–184, 206
Microsoft Network, 86
Microsoft Windows, 224, 226
Middlemen. *See* Disintermediation
Miller, Arthur, 133, 134
Milosevic, Slobodan, 7, 9
Misinformation. *See* Information
MSNBC, 86
Myanmar, 50, 66
Myrhvold, Nathan, 88, 96
MySki, 47–48

Nader, Ralph, 95
Narrowcasting, 46
National Governors' Association, 149
National Inquirer, 190
National Public Radio, 190
National Public Telecommunications
    Network, 210
National Research Council, 77
National Rifle Association (NRA), 13–14
National Science Foundation (NSF),
    20, 185
National Security Agency, 75–76
Negroponte, Nicholas, 45, 105–106, 118
Netherlands, 22
Netscape, 44, 91, 92, 173, 183
Networks
    broadband, 17
    community, 212–215. *See also* Com-
        munities
    and computers, 92
    distributed, 16
    effects, 93–94
    packet-based, 16
    television, 16, 40
    See also Internet
"New context" approach, 170, 172
New Republic, 141
News media. *See* Journalism
News personalization services, 45–47,
    105–106, 113

Newspage, 45
Newsweek, 135
New York Islanders, 208
New York Times, xvi, 47, 98
Norway, 22
Novak, Tom, 36, 37
NRA. *See* National Rifle Association
NSF. *See* National Science Foundation

Online communities, 48–49, 120–122
Online information, 133–141
Online investing, 53–56, 145–147
Open-mindedness, 200–201
Open-source software, 184–185
Operating systems, 28, 88–91, 94–95, 226
Order and chaos, 197–207
Ornstein, Norman, 134
"Oversteering," 104
Ownership of media. *See* Mass media, ownership

P3P. *See* Platform for Privacy Preferences Project
Packet-based networks 16
Paine, Thomas, 5
Pantic, Drazen, 7
ParoleWatch, 150–153
Pay-to-be-heard model, 130–131
PCs. *See* Personal computers
Perot, Ross, 58
Personal computers (PCs), 26–27, 88, 93–95
Personal control. *See* Control
Personalization, 44–52, 105–114, 198–199
PICS. *See* Platform for Internet Content Selection
Platform for Internet Content Selection (PICS), 108
Platform for Privacy Preferences Project (P3P), 160–161
"Platform independent" technology, 91, 183
Points of access, 18
Political activists. *See* Activism
Politics, 57–59, 150–158, 192–196
Pool, Ithiel de Sola, 123, 217, 226
Poor, 162–163

Pornography, 34–35, 68–70. *See also* Censorship
Portal sites, 99
Post, David, 36
PriceScan, 56
"Principles in context" approach, 170, 172, 176, 177, 228
Printing history, 5–6
Privacy, 51, 158–165, 175–177, 219–221
Privatization of Internet, 20–21
Protocols, 17, 73, 184
Psychology of control, 22–24
Public domain, 78–79, 178–179
"Public forum" doctrine, 125, 203
Public libraries, 171
PublicNet, 205–206, 215, 227
Push technology, 99–100, 182

Radikal, 69
Radin, Margaret Jane, 164
Radio B92, 7–8, 15, 49, 55, 124
Randomness, 198–207
Rashomon, 189
Reed, Ralph, 115, 117
Rees-Mogg, William, 229
Release 2.0 (Dyson), 121
Remote communications, 193–194
Reno, Janet, 90
Renshon, Stanley 22–23
Representative democracy. *See* Democracy
Resistance to control revolution, 61–101
Revolution, 9–11
Rheingold, Howard, 41, 211–212
Rifkin, Jeremy, 12
Rimm, Marty, 35, 37, 38
The Road Ahead (Gates), 25
Roberts, Cokie, 195
Roberts, Steve, 195
Rockefeller, John D., 90
Roszak, Theodore, 119
Rules for the Internet, 169–179

Safdar, Shabbir, 51
Safety nets, 219–221
Sales taxes, 147–149, 222–223
Schools. *See* Education

Sclove, Richard, 143
Screen bias, 95–97
Screening. *See* Filtering
Search engines, 95–96, 98–99
SEC. *See* Securities and Exchange Commission
Second International Harvard Conference on Internet and Society, 133
Secret information, 73–78
Securities and Exchange Commission (SEC), 145
Seiger, Jonah, 51
Selective avoidance, 109–111
Serbia, 7–8
Serendipity. *See* Randomness
Sexual materials, 170–174. *See also* Pornography
Shaw, David, 118
Shenk, David, 44–45
Shortwave radios, 6
Sidewalk (Microsoft), 86
Singapore, 66
Slate, 86, 190
Social activists. *See* Activism
Social fragmentation, 119–120
Software, 86, 113, 227
    browsers, 90–91, 173–174, 182–183
    bundled, 182
    indecency-blocking, 172
    open-source, 184–185
Somm, Felix, 68–69
South Korea, 66
Soviet Union, 3–5, 8
    mass media, 6
    nuclear attack, 20
Spain, 22
Speech, 64–73, 108, 124–132, 203–207
Springsteen, Bruce, 41
Stanley Cup, 208
Star Trek, 114
Starr, Kenneth, 135
Stephanopolous, George, 135
Sun Microsystems, 183
Sunstein, Cass, 108–109
Supreme Court, 38, 70, 125, 128, 172, 198, 204, 207
Sweden, 22

Taxes, 147–149, 222–223
TCP/IP. *See* Transfer Control Protocol/Internet Protocol
Technologies of Freedom (Pool), 217
Technology
    and children, 202–203
    and communications. *See* Communications
    impact of, 13–15
    platform independent, 183
    push, 99–100, 182
    See also Internet
Telecommunications Act of 1996, 19–20
Telecommuting, 48, 112
Teledemocracy, 59, 195. *See also* Democracy
Telephones, 41–42
Television
    as babysitter, 18
    networks, 16, 40
    Serbia, 7
Tiananmen Square, 6
Tierney, John, 11
Time (Magazine), 34–37, 38, 55, 124, 141, 189
Today's Papers, 190
Total filtering. *See* Filtering
Transfer Control Protocol/Internet Protocol (TCP/IP), 17
TRUSTe, 160
Trusted systems, 81–82, 178, 179
"Truth watchers," 189–190
Turkle, Sherry, 25

Universal access, 17–18, 224
Usenet discussion groups, 42

VHS standard, 94
Vietnam, 66
Virtual communities. *See* Online communities
The Virtual Community (Rheingold), 211
Virtual Vineyards, 56, 146
Visual artists, 42–43
Vonnegut, Kurt, 139
Voter registration, 194–195
Voters Telecommunications Watch, 51

Wall Street Journal, 138
Washington Post, 47
Watermarks, 80
Web sites
    commercial, 214–215
    design of, 184
    newspaper, 46–47
WebTV, 87, 99–100, 180
Well (BBS), 35–36, 211
Wiener, Norbert, 32
Windows operating systems. *See* Microsoft Windows; Operating systems

WIPO. *See* World Intellectual Property
    Organization
Wired, 188
Wiretap, 175–176
Word processing programs, 27–28, 183
World Intellectual Property Organization (WIPO), 80
World Trade Organization, 228

Y2K problem, 228
Yahoo, 44, 98, 100
Yugoslavia, 7–8

PUBLICAFFAIRS is a new nonfiction publishing house and a tribute to the standards, values, and flair of three persons who have served as mentors to countless reporters, writers, editors, and book people of all kinds, including me.

I. F. STONE, proprietor of *I. F. Stone's Weekly*, combined a commitment to the First Amendment with entrepreneurial zeal and reporting skill and became one of the great independent journalists in American history. At the age of eighty, Izzy published *The Trial of Socrates*, which was a national bestseller. He wrote the book after he taught himself ancient Greek.

BENJAMIN C. BRADLEE was for nearly thirty years the charismatic editorial leader of *The Washington Post*. It was Ben who gave the *Post* the range and courage to pursue such historic issues as Watergate. He supported his reporters with a tenacity that made them fearless, and it is no accident that so many became authors of influential, best-selling books.

ROBERT L. BERNSTEIN, the chief executive of Random House for more than a quarter century, guided one of the nation's premier publishing houses. Bob was personally responsible for many books of political dissent and argument that challenged tyranny around the globe. He is also the founder and was the longtime chair of Human Rights Watch, one of the most respected human rights organizations in the world.

. . .

For fifty years, the banner of Public Affairs Press was carried by its owner Morris B. Schnapper, who published Gandhi, Nasser, Toynbee, Truman, and about 1,500 other authors. In 1983 Schnapper was described by *The Washington Post* as "a redoubtable gadfly." His legacy will endure in the books to come.

Peter Osnos, *Publisher*